UNDERSTANDING HYDROLOGICAL PROCESSES IN AN UNGAUGED CATCHMENT IN SUB-SAHARAN AFRICA

Understanding Hydrological Processes in an Ungauged Catchment in sub-Saharan Africa

DISSERTATION

Submitted in fulfilment of the requirements of
the Board for Doctorates of Delft University of Technology
and of the Academic Board of the UNESCO-IHE Institute for Water Education
for the Degree of DOCTOR
to be defended in public
on Friday, 6 February 2009 at 12:30 hours
in Delft, The Netherlands

by

Margaretha Louise MUL
born in Heinenoord, the Netherlands
Master of Science of Civil Engineering, Delft University of Technology

This dissertation has been approved by the supervisors
Prof.dr.ir. H.H.G. Savenije   TU Delft/ UNESCO-IHE, The Netherlands
Prof.dr. S. Uhlenbrook   UNESCO-IHE/ VU Amsterdam, The Netherlands

Members of the Awarding Committee:

Chairman:	Rector Magnificus, TU Delft, The Netherlands
Vice-chairman:	Rector, UNESCO-IHE, The Netherlands
Prof.dr.ir. H.H.G. Savenije	TU Delft/ UNESCO-IHE, The Netherlands, supervisor
Prof.dr. S. Uhlenbrook	UNESCO-IHE/ VU Amsterdam, The Netherlands, supervisor
Prof.dr.ir. N.C. van de Giesen	TU Delft, The Netherlands
Prof.dr. A. Bronstert	U-Potsdam, Germany
Prof.dr. E. Zehe	TU München, Germany
Prof.dr.ir. T.N. Olsthoorn	TU Delft, The Netherlands
Prof.dr.ir. P. van der Zaag	UNESCO-IHE/ TU Delft, The Netherlands

CRC Press/Balkema is an imprint of the Taylor & Francis Group, an informa business

Published by:
CRC Press/Balkema
PO Box 447, 2300 AK Leiden, The Netherlands
e-mail: Pub.NL@taylorandfrancis.com
www.crcpress.com – www.taylorandfrancis.co.uk – www.balkema.nl

ISBN 978-0-415-54956-1 (Taylor & Francis Group)

ABSTRACT

Ungauged catchments can be found in many parts of the world, but particularly in sub-Saharan Africa. Information collected in a gauged catchment and its regionalisation to ungauged areas is crucial for water resources assessment. Especially farmers in semi-arid areas are in need of such information. Inter and intra-seasonal rainfall variability is large in these areas, and farmers depend more and more on additional surface and groundwater resources for their crop production. As a result, understanding of the key hydrological processes, and determination of the frequencies and magnitudes of stream flows, is very important for local food production. This is particularly true for the ungauged Makanya catchment in Tanzania, which is the subject of this study.

In the absence of long-term hydrological data, hydrological processes have been studied through a multi-method approach. Regular rainfall and runoff measurement devices were installed in a nested catchment approach. High spatial and temporal resolution data have been collected over a period of 2 years to capture all the hydrological processes. Spring samples have been taken to identify groundwater flow systems. Hydrograph separation with hydro-chemical data has been performed to identify and quantify the origins and flow pathways of the water during flood flows. Electrical resistivity tomography (ERT) has been used to map the subsurface structure at selected sites. Finally, a conceptual model has been developed to test the hypothesised conceptualisation of the flow paths.

Agricultural practices by farmers in the catchment vary as a function of location as they are influenced by the local climate, water resources availability and soil type. Three zones with distinct features have been identified within the study area. In the highlands, it is cooler, rainfall is more abundant and there are perennial springs. Here irrigation is practiced as supplementary irrigation in the wet season and as full irrigation during the dry season. In the midlands only supplementary irrigation is practiced using the remainder of the perennial streams coming from the highlands. Here the high probability of occurrence of dry spells requires supplementary irrigation. In the lowlands, base flow has dried up and spate-irrigation is practiced during the rainy season, whereby flash floods are diverted from the main river onto the farm land.

In the highlands, the occurrence of perennial springs is defined by the geology. These springs discharge a substantial amount of water. In the midlands, few springs exist, yielding a substantially lower amount of discharge, with poorer quality. Abundance of spring water from the highlands is both used in the highlands and the midlands for irrigation. However, due to the many diversions, the rivers do no longer reach the outlet of the catchment, as they do not exceed the infiltration capacity of the alluvium. Only large flash floods reach the spate-irrigation system in the lowlands.

Two types of floods have been analysed in the catchment, induced by different types of rainfall. High intensity rainfall can generate flash floods which reach the spate-irrigation system. Smaller floods infiltrate into the alluvium before reaching the spate-irrigation system. It has been observed that during small floods, the vast majority of the flood originates from groundwater (> 90 percent). During an extreme event, the groundwater contribution is still about 50 percent of the total flow, increasing the outflow from the groundwater reservoir a 100-fold during the peak flow and causing a substantial base flow increase after the event. The time of concentration is extremely short, within one to two hours the flood peak reaches the outlet of the catchment flowing into the spate-irrigation system. The rainfall during the March 2006 event was such an extreme event, with intensities as high as 50 mm hr^{-1}, while the entire event lasted only 4-5 hours. During this event, rainfall intensities exceeded the infiltration capacity of the soils and overland flow was common. The flows exceeded also the absorption capacity of the alluvium of the main valley of the catchment, replenishing the local aquifer and feeding the downstream spate-irrigation system.

The temporal resolution of the observed data is 15 min. This level of detail was necessary to capture the rapid catchment response during peak flows. The observation data, unfortunately is not complete as extreme events damaged the structures and affected the reliability of the observations. Moreover, 7 out of 10 instruments were lost, stolen or damaged which hampered the data collection.

A conceptual hydrological model has been developed to test the rainfall-runoff hypotheses. The model is able to model in a process-based fashion the flows at the foot of the mountain, incorporating the observed hydrological processes. The model yields good results for a simulation of a 2-year period at hourly time step (Nash-Sutcliffe efficiency: 0.79 and Log Nash-Sutcliffe: 0.90). This model contains a large number of parameters, of which several parameters could be identified from the data. However, automatic optimisation of the remaining parameters is hampered by equifinality. Hence, the strategy chosen is step-wise calibration using multiple performance criteria judging hydrograph performance visually. If the hydrological model would be used for upscaling farming practices, then one should realise that the hydrological processes in the valley of the catchment are different from the highland processes and need to be studied in further detail before the model can be upscaled to the catchment level.

The current hydrological and water resources situation, as in many parts of Africa, is also a result of anthropogenic influences. Increased water usage in the upper parts of the catchment generated the need for agreements between highland and midland water users. No agreements are in existence between these two groups and the lowland farmers. With the base flow no longer reaching Makanya, the farmers were forced to change their irrigation practices. Changes in land use in the upstream parts have also impacted on the hydrological processes. However, with a lack of historical data, we were unable to quantify this. The downstream farmers are not solely negatively affected, as flash floods generated in the highlands reach Makanya, replenish the unsaturated zone and local groundwater bodies and deposit fertile sediments. Currently, there is a balance between the upstream and downstream farmers, whereby yields are produced by each system. However, it is unsure how future increases in water requirements will affect this balance.

ACKNOWLEDGEMENT

This thesis is part of the Smallholder System Innovations in Integrated Watershed Management (SSI) Programme funded by the Netherlands Foundation for the Advancement of Tropical Research (WOTRO), the Swedish International Development Cooperation Agency (Sida), the Netherlands Directorate-General of Development Cooperation (DGIS) and the International Water Management Institute (IWMI). The on-site implementation was facilitated by the Soil-Water Management Research Group (SWMRG) at Sokoine University of Agriculture, Tanzania. Moreover I would like to say special thanks to UNESCO-IHE Institute for Water Education for their financial support.

Since I started my research almost 5 years ago, I met many people, who helped me with my PhD research and made me feel at home in their countries. In the first place I would like to recognise my colleagues within the SSI team for their companionship and valuable cooperation; together we managed to carry out a great idea into practice. Although I can not name all of you, I would like to mention a few in particular; Elin for your support and friendship while we were in the; Hans and Jeltsje for working close with me in the field for the past two years and most of all Hodson Makurira for all the support, friendship, stories and laughs that we shared in Tanzania, Zimbabwe and the Netherlands.

I would also like to thank my supervisors prof. Savenije and prof. Uhlenbrook for all the constructive discussions and useful advice. I am also appreciative of all the support I received from my colleagues at UNESCO-IHE, especially the MAI department and the laboratory who helped me a lot with the many water samples that needed to be analysed.

I would also like to thank the MSc students from various universities, whom I worked closely with and whose research outputs contributed to this thesis: Festo Nchunguye, Mbonea Mshana and Justin Mutiro from the University of Zimbabwe; Robert Mutiibwa, from UNESCO-IHE; Lennart Woltering, Maurits Voogt, Benjamin Fischer and Ronald Bohté, from Delft University of Technology; Olivier Faber from Freiburg University and Timo Kessler from Stuttgart University.

I also would like to express my thanks to my colleagues Hoko, Misi, Mhizha, Siwadi, Kaseke and Makurira at the University of Zimbabwe who welcomed me and made me feel at home in Harare. I also would like to thank Lewis, Themba, Martha, Nick, Joanna, Admire, Dorcas, Nyamuzwera and Moriah from the WaterNet Secretariat for letting me use your office supplies and all the "pa-tea" times in the morning.

In Makanya, Tanzania, I would like to thank all the data collectors who collected the valuable data presented in this thesis and protected the delicate equipment. Special thanks to our driver Hamadi Salum for taking us safely from place to place, and Istamil Msangi, Maliki Abdallah and Ally Hussein for assisting us in the field. With the long days we spent in the field, there is a need for food and refreshments. Therefore, I will always remember the girls at Mama Ntilie for the nice food they prepared for us and our house in Makanya owned by Mama Shabani.

Finally, I would like to thank all the PhD students at UNESCO-IHE and TU-Delft for their friendship and nice discussions. Last but not least I would like to thank my family and friends for supporting me even while I was far away.

PREFACE

My PhD research started in 2003, when prof. Savenije asked me if I wanted to do a PhD in Tanzania and live in Zimbabwe. I immediately accepted because the subject was relevant and fitted with my hydrology background. Although at times, I thought "What have I gotten myself into", I never regretted this decision. My research topic focussed on assessing the impact of smallholder system innovations on downstream water availability. The challenge was to first understand the hydrological processes in the study area, the Makanya catchment in Tanzania, before the impact of smallholder system innovations could be assessed. The Makanya catchment was completely ungauged and required intensive fieldwork. This consisted of installing on-site equipment to collect rainfall and runoff data which was a time consuming process that took two years before the first complete set of data was available. This was due to a number of problems; some equipment was installed at the wrong place due to a limited knowledge of the areas characteristics, some equipment was stolen and some was damaged by unexpected large floods. Therefore, I had to regularly check and repair the measuring structures. The data derived from this fieldwork was used to understand the hydrological processes underlying water stresses that farmers face in the catchment. I hope the data collection will continue in this area, as it would be a pity if all the efforts made in setting up and maintaining this measuring network were lost.

Due to the high climatic variability in the area, the collection of hydrological data was very difficult. Since the spatial variability requires a dense network of measuring instruments, I installed 42 raingauges in the 300 km^2 catchment area, and I measured runoff at 5 different locations. The quick response time of the catchment also requires high temporal resolution of the data, requiring automated equipment. Extreme events hampered the data collection, by destroying discharge gauges and pressure transducers. Of the 4 years we collected data, one year was exceptionally dry (2005) and one extremely wet (2006) coinciding with an El Niño season. This shows that there is no typical average year which can be used to generalise the situation for this study area. Normally, a longer data period is needed to assess the climatic variability. However, because of the high spatial and temporal resolution of the data collected, it was possible to reach conclusive results.

The original objective of the study, which was to assess the impact of farmer innovations on the hydrology, could not be met for different reasons including the lack of existing data, the time required for the new data collection and the subsequent analyses to understand the hydrological processes in the study area. Moreover at the outlet of the catchment, the river is intermittent only flowing during extreme events. Only the mountain streams could be monitored continuously whereas the focus of the SSI programme was in the valley of the catchment. This meant that the research done by other SSI team members could not be used to assess the impact of upstream developments. This study, therefore, only focused on the understanding of the hydrological processes in the mountain.

Besides learning much about hydrological processes in a semi-arid catchment I gained enourmous personal experience by spending a lot of time in the field where I got to know a large variety of people, where I made many friends and came to appreciate different ways of life. I believe this experience has enriched my life and left me with many friends from all over the world.

LIST OF SYMBOLS

Symbol	Parameter description	value/ unit	dimension
α	Climate dependent factor (Allen et al., 1998a)	1.35	-
A_f	Flow through area		L^2
A	Catchment area		L^2
β	Climate dependent factor (Allen et al., 1998)	-0.35	-
b	Bottom width		L
B	Width of the cross section		L
B_{LEW}	coefficient for determining the spatial variability of the soil moisture capacity		-
c	Climate dependent factor (Allen et al., 1998)	0.34	-
c_p	Specific heat of dry air at constant pressure (Allen et al., 1998)	1004.6 J kg^{-1} K^{-1}	
C_a	Concentration of anion		M L^{-3}
C_c	Concentration of cation		M L^{-3}
c_T	Concentration at sampling point		L^3 T^{-1}
c_S	Concentration of surface runoff		L^3 T^{-1}
c_G	Concentration of sub-surface runoff		L^3 T^{-1}
d	Climate dependent factor (Shuttleworth, 1993)	-0.14	-
D_r	Root depth		L
D	Threshold value for interception		L
E	Total Evaporation		L T^{-1}
E_I	Interception		L T^{-1}
E_T	Transpiration		L T^{-1}
E_s	Evaporation from the soil		L T^{-1}
E_{pot}	Potential open water evaporation		L T^{-1}
$E_{t,\,pot}$	Potential transpiration		L T^{-1}
E_{pan}	Pan evaporation		L T^{-1}
e_a	Vapour pressure at temperature T_a		M L T^{-2}
e_d	Prevailing vapour pressure at the evaporating surface		M L T^{-2}
e_s	Vapour pressure		M L T^{-2}
f	Splitting fraction		-
F	Infiltration		L T^{-1}
F_{LEW}	fraction of the LEW that has a lower soil moisture capacity that S_u		-
G	Heat flux density into the water body		M T^{-1}
g	Gravity	9.81 m s^{-1}	
h	Humidity		-
$\partial h/\partial t$	Change in head over time		L T^{-1}

Symbol	Parameter description	value/ unit	dimension
i_e	Energy slope		-
i_b	Bed slope		-
I_b	Ion balance		-
λ	Latent heat coefficient (Shuttleworth, 1993)	2,450,000 J kg^{-1}	
ρ	Density of water (Shuttleworth, 1993)	1000 kg m^{-3}	M L^{-3}
ρ_a	Density of moist air		M T^{-3}
γ	Psychrometric constant (Shuttleworth, 1993)	0.067 kPa $^{\circ}$C^{-1}	
σ	Stefan-Boltzmann constant (Shuttleworth, 1993)	2.04*10^{-4} J m^{-2} K^{-4} d^{-1}	
K_p	Pan factor	0.7	-
K_c	Crop coefficient		-
K	Time scale		T
K_{q1}	Time scale, direct runoff		T
K_q	Time scale, quick flow		T
K_s	Time scale, slow flow		T
K_m	Time scale, groundwater leakage to Mbaga		T
$K_{v,a}$	Time scale, groundwater leakage to Vudee		T
K_v	Time scale, groundwater connection to Vudee		T
l	Water level		L
L	Gradually varied flow parameter	3.33	-
M	Gradually varied flow parameter	3	-
n	Measured solar radiation		M T^{-3}
n_m	Roughness coefficient		T L$^{-1/3}$
N	Max observed solar radiation		M T^{-3}
n_c	Critical dry spell length		T
P	Precipitation		L T^{-1}
P_n	Net precipitation		L T^{-1}
P_{e1}	Excess rainfall component 1		L T^{-1}
P_{e2}	Excess rainfall component 2		L T^{-1}
p_{01}	Transition probability of a rainday after a dry day		-
p_{11}	Transition probability of a rainday after a rainday		-
θ	Soil moisture retention capacity of the soil		-
Q_{obs}	Observed discharge		L^3 T^{-1}
Q_{sim}	Simulated discharge		L^3 T^{-1}
Q	Total discharge		L^3 T^{-1}
Q_s	Direct surface runoff		L^3 T^{-1}
Q_u	Runoff through the unsaturated zone		L^3 T^{-1}
Q_g	Groundwater flow		L^3 T^{-1}
Q_{max}	Peak discharge		L^3 T^{-1}
Q_{est}	Estimated discharge		L^3 T^{-1}
R	Recharge to the groundwater		L T^{-1}

Symbol	Parameter description	value/ unit	dimension
R_h	Hydraulic radius		L
r	Albedo		-
R_n	Energy flux density of the net incoming radiation		M T^{-3}
R_{ns}	Net incoming short wave radiation		M T^{-3}
R_s	Observed incoming radiation		M T^{-3}
R_{nl}	Net outgoing long wave radiation		M T^{-3}
r_a	Aerodynamic diffusion resistance		T L^{-1}
r_s	Canopy diffusion resistance		T L^{-1}
s	Slope of the saturated vapour pressure curve		M L Θ T^{-2}
S	Storage		L^3
S_u	Storage unsaturated zone		L^3
$S_{u,max}$	Maximum storage of unsaturated zone		L^3
S_s	Storage saturated zone		L^3
$S_{s,\,th}$	Threshold value for quick runoff		L^3
$S_{s,\,max}$	Threshold value for direct runoff		L^3
$S_{m,th}$	Threshold value connecting to Vudee		L^3
$\dfrac{dS}{dt}$	Change in storage of the catchment		L T^{-1}
T_a	Actual temperature		Θ
t	Spearman rank test		-
u_2	Wind speed at 2m		L T^{-1}
y_N	Normal depth		L
y_c	Critical depth		L
y	Actual water depth		L
dy/dx	Change of water depth		-
z	Slope of the banks		-

LIST OF ACRONYMS

EC	Electrical Conductivity
ERT	Electrical Resistivity Tomography
HYMOD	
IUCN	International Union for Conservation of Nature
JICA	Japan International Cooperation Agency
LEW	Lumped Elementary Watershed
NGO	Non-governmental organisation
PBWO	Pangani Basin Water Office
SAIPRO	Same Agricultural Improvement Program
SSI	Smallholder systems innovations in Integrated Watershed Management
SUA	Sokoine University of Agriculture
TIP	Traditional Irrigation Improvement Project
UNYINDO	Water user group association between Bangalala and Ndolwa
URT	United Republic of Tanzania
WUG	Water user groups

TABLE OF CONTENT

Abstract   v

Acknowledgement   vii

Preface   viii

List of Symbols   ix

List of Acronyms   xii

Table of Content   xiii

Chapter 1 Introduction   1

1.1   Background-- 1
1.2   Impact of Land Use and Management--------------------------------------- 2
1.3   Hydrological Process Understanding-------------------------------------- 6
1.4   Hydrological Modelling-- 8
1.5   Objectives--10

Chapter 2 Study Area   11

2.1   Pangani River Basin --- 11
2.2   South Pare Mountains--- 13
      2.2.1 Geology... 14
      2.2.2 Rainfall.. 15
      2.2.3 Agricultural water use 20
      2.2.4 Runoff.. 20

Chapter 3 Weather and Climate   23

3.1   Rainfall-- 23
3.2   Spatial Rainfall Variability --- 32
3.3   Potential Evaporation -- 35
3.4   Potential Transpiration versus Rainfall------------------------------- 38

Chapter 4 Investigation of Flow Systems   39

4.1   Hydrochemical Mapping -- 39
      4.1.1 Water Quality of Springs 40
      4.1.2 Comparison of Dry versus Wet Conditions 43
4.2   Geophysical Investigations-- 45
      4.2.1 Site I: Highlands... 46
      4.2.2 Site II: Kilenga Spring....................................... 46
      4.2.3 Site III: Valley Cross Section................................ 48
      4.2.4 Site IV: Stream Bed... 49
      4.2.5 Auger Wells 49
4.3   Spring Types-- 50
4.4   Synthesis: Groundwater Flow Systems in the Makanya Catchment---------- 51

4.5   Base flow fluctuations-- 52

     4.5.1 Observations 53

4.6   Conclusion -- 57

Chapter 5 Investigation of Hydrological Events   **59**

5.1   Hydrograph Separation using Hydrochemical Tracers -------------------------- 59

     5.1.1 9 November Event 60

     5.1.2 5 December Event.. 64

     5.1.3 Discussion.. 67

5.2   Spatial Rainfall Variability and Runoff Response during an Extreme Event 68

     5.2.1 Rainfall.. 68

     5.2.2 Runoff... 70

     5.2.3 Water Quality... 76

     5.2.4 Conclusions.. 78

Chapter 6 Hydrological Modelling   **81**

6.1   Observations --- 81

6.2   Methodology --- 86

6.3   Model Description --- 87

6.4   Optimisation --- 89

6.5   Results--- 90

6.6   Conclusions-- 92

Chapter 7 Sharing Water   **95**

7.1   Water Allocation Practices-- 95

7.2   agricultural Water Users in Makanya Catchment---------------------------- 96

     7.2.1 Manoo Furrow System... 97

     7.2.2 Makanya Spate-irrigation System 102

7.3   Water Sharing between Users of Adjacent Furrows -------------------------104

7.4   Water Allocation between Neighbouring Villages-------------------------105

     7.4.1 Agreement between Ndolwa and Bangalala 105

     7.4.2 Agreement between Vudee and Bangalala 106

7.5   Water Sharing between Distant Villages ----------------------------------107

7.6   Discussion and Concluding Remarks--109

Chapter 8 Syntheses and Conclusions   **111**

8.1   Farming Adaptation to Climatological, Hydrological and Bio-physical

Constraints --111

8.2   Hydrological Process Understanding and Modelling ----------------------112

8.3   Impact of farming activities on hydrology--------------------------------114

References   **117**

Samenvatting   **127**

About the Author   **130**

Chapter 1

INTRODUCTION

1.1 BACKGROUND

Global population drives the water requirements for domestic use and in particular food, which determines close to 90 percent of the total water requirement of a person (Savenije, 2000). With changing diets the water requirement may even become much higher (with meat requiring even more water (Hoekstra and Chapagain, 2007; Liu and Savenije, 2008). If we compare the water requirements and the water resources of a country, many countries in sub-Saharan Africa are considered water scarce. However, such analysis considers only "blue" water and not "green" water (Savenije, 2000). "Blue water" is the water which occurs in aquifers, rivers and lakes. In sub-Saharan Africa, 70-90 percent of the exploited "blue water" is used by irrigated agriculture. Increasing the agricultural production by increasing the irrigated areas is becoming more and more difficult, whereby suitable areas are becoming less available and "blue water" resources are diminishing. However, ninety percent of the agricultural land in sub-Saharan Africa is rainfed (Rockström, 2000), which, in fact, uses "green water" (Falkenmark, 1995). "Green water" is the water that is returned to the atmosphere through transpiration and is the most under-valued resource (Savenije, 2000). Improving the productivity of rainfed agriculture in sub-Saharan Africa appears the best strategy. However, using this "green water" for agriculture may have an impact on the available "blue water" resources, by changing rainfall partitioning.

The programme for Smallholder Systems Innovations in Integrated Watershed Management (SSI) is an integrated research programme which aims to develop strategies for sustainable solutions to improving rainfed agriculture (Bhatt et al., 2006). One of the research focuses is the hydrological implications of implementing these innovations in a semi-arid environment. These techniques can range from *in situ* practices such as deep tillage and zero tillage to infrastructural interventions such as underground storage tanks and small runoff harvesting structures. They were studied at field scale ranging from water balance studies (Kosgei et al., 2007; Makurira et al., 2008b) to slow-changing soil variables (Bhatt et al., 2006). In general, at field scale the interventions alter the rainfall partitioning, directing more water to productive use (transpiration) and promoting infiltration, reducing direct runoff and soil evaporation. However, the question how these changes translate to large scale impacts is still a relatively unknown territory. This research is focussing on the large scale hydrological implications of these techniques. In order to understand these, the processes occurring in the catchment have to be understood at the different scales, before the impact of these innovations can be assessed.

Understanding the changes in the hydrological regime is important because it will affect downstream water availability, water users and water using activities. Without knowing the downstream consequences, no holistic approach can be applied for the management of the catchment. When managing a system, the question is raised how to balance the available water for all users with the emphasis on water for food production and ecosystems. With the knowledge on downstream consequences of certain land use and management options, well informed decision can be made on catchment planning. While the aim is to increase food production, other stakeholders in the catchment should not be neglected. The challenge is to identify to what extent technologies can be implemented without jeopardising other uses and functions downstream.

1.2   IMPACT OF LAND USE AND MANAGEMENT

Population growth, in general, induces land use change, using more land for crop production and in the same process, reducing the forested land or land occupied by natural ecosystems. In addition, the migration of people from the rural to the urban areas changes land into urban land use. Although these changes do not use water directly, they do have a consequence on the hydrology. Changes in transpiration and infiltration rates have an impact on runoff generation and recharge. Similarly, land management, such as soil and water conservation measures in agriculture, can affect rainfall partitioning (Lørup et al., 1998). Often identifying land use change is done by remote sensing techniques, which are not always capable of observing land management changes.

The changes in hydrological regime due to land use or management changes are related to the two partitioning points as illustrated by Fig. 1.1. The first partitioning point determines the effective rainfall, which equals the rainfall (P) minus the evaporation from interception (E_I) and divides it into direct surface runoff (Q_s) and infiltration (F). Governing processes are the infiltration rate (determining the actual infiltration) and interception, depending on vegetation cover and soil surface roughness. The second partitioning point separates the infiltrated water into transpiration (E_T), evaporation from the soil (E_s) and recharge to the groundwater (R). Water stored in the unsaturated zone can also come to runoff through the unsaturated zone (Q_u). Vegetation cover and crop stage influence the partitioning at this partitioning point.

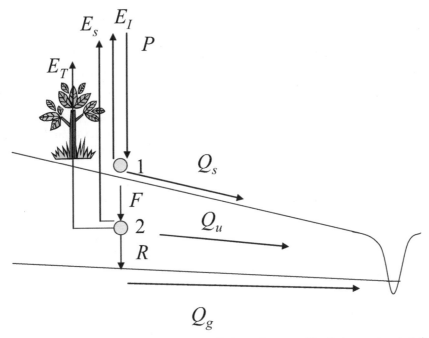

Figure 1.1 The two rainfall partitioning points, Q_s is surface runoff, Q_u is unsaturated flow, Q_g is groundwater flow, P is precipitation, E_s is evaporation from interception, E_s is evaporation from the soil, E_T is transpiration, F is infiltration, and R is recharge to the aquifer.

Transpiration rates are different for different vegetation covers, whereby forests generally transpire at higher rates than shrubs or agricultural lands. Calder (1999) showed that for a deforested catchment, in general, the total annual runoff increases as a result of decreasing transpiration (most of these studies were done in humid climates). This is shown by the water balance equation 1.1:

$$\frac{\mathrm{d}S}{\mathrm{d}t} = (P - E) - Q\!\!\big/\!\!A \qquad\qquad\qquad\qquad \text{eq. 1.1}$$

where:

$\dfrac{\mathrm{d}S}{\mathrm{d}t}$ = change in storage of the catchment [L T^{-1}],

P = precipitation in the catchment [L T^{-1}],

A = catchment area [L^2],

Q = discharge from the catchment [L^3 T^{-1}], which consists of:

$$Q = Q_s + Q_u + Q_g \qquad\qquad\qquad\qquad \text{eq. 1.2}$$

E = total evaporation [L T^{-1}], which consists of:

$$E = E_I + E_s + E_T \qquad\qquad\qquad\qquad \text{eq. 1.3}$$

On long term average, the change in storage can be considered negligible ($\frac{dS}{dt} = 0$). The equation then becomes:

$$\overline{Q} \cong \overline{P} - \overline{E} \qquad\qquad\qquad \text{eq. 1.4}$$

where the over bar indicates annual average values.

This, however, doesn't take into account the intra-seasonal variability, which is particularly apparent in tropical regions. Seasonal rainfall variability, with clear wet and dry season affects the impact of land use and management of a catchment in a different way then if there is uniform rainfall.

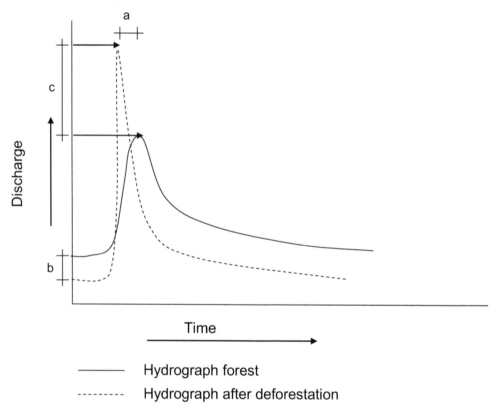

Figure 1.2 Conceptual representation of changes in hydrograph, a) reduced time of concentration and b) reduced base flow, and c) higher peak flows.

The annual water balance may be the same after a change in land use but it does not necessarily mean that at a smaller time scale there is no change. Fig. 1.2 shows a hypothetical case for a single hydrograph. It shows a forested and cleared catchment, the deforested catchment has a smaller time of concentration (a) and peak flows are larger (c) compared to the forested catchment (Brown et al., 2005). Base flow on the other hand is smaller (b), reduced infiltration rates reduce the replenishment of the groundwater aquifer and subsequently the base flow. In addition, the peak of the flow may increase, which could have severe consequences downstream. However, the volume under the hydrograph may change in either direction, indicating an increase or decrease of the total runoff. Land use change can direct the runoff response of a catchment in different ways, depending on the rainfall partitioning. The processes governing rainfall partitioning have a time scale ranging from one day (interception) to a couple of weeks (transpiration) after the actual rainfall took place (Savenije, 2004), depending on rainfall intensity, soil conditions and cover. However, the implications for the hydrological regime of the catchment have a longer time scale depending on the spatial scale. For example, the difference in residence time of a particle that infiltrates and recharges the groundwater compared to a particle that becomes surface runoff is from a day to up to years. That means that if the path of a particle is changed it will affect the residence time and hence the runoff (see also Fig. 1.2). Not only can the changes in rainfall partitioning affect the total annual runoff, it, more so, affects the inter-annual distribution of the runoff. The actual impact of the land use change is dependent on the local conditions, from local climate to hydrological processes.

Several techniques have been used to assess the runoff response from different land uses. Brown et al. (2005) used the paired catchment approach whereby the runoff response of two neighbouring catchments with different land uses are analysed. Calder et al. (1995) used time series analysis of a catchment, including the "old" situation and the "new" situation. Differences in the runoff response are then attributed to the different land uses. The complication with using this method is that changes in the climate can not be filtered out of the analyses and a long time series is necessary. Hydrological modelling is also often used to identify the possible changes in the runoff response of a catchment (e.g. Fenicia et al., 2008; Ott and Uhlenbrook, 2004).

Results from paired catchment studies, whereby two similar and nearby catchments are monitored, have generated a large amount of information (Brown et al., 2005; Sahin and Hall, 1996), particularly on small scale hydrological implications (1ha - 10km^2). Ideally the catchments are initially monitored with similar land use and a correlation established. Afterwards for one of the catchments the land use is changed, e.g. the trees are harvested, or a forest is changed towards agricultural or barren land. Many studies have used this methodology to assess the impact of land use changes on the hydrology (with a summary presented by Brown et al. (2005); Bosch and Hewlett (1982)). One of the main conclusions, which are relevant to this study, is that the major impacts in semi-arid and arid climates are observed during base flow, especially for areas with distinct rain and dry seasons (Brown et al., 2005; Bruijnzeel, 1988; Edwards, 1979; Scott et al., 2000). Additionally, it has been observed that clearing of riparian vegetation increases the flow in the river in some selected South African catchments (Dye and Poulter, 1995; Prinsloo and Scott, 1999). This is particularly important during periods of low flows and in case of invasive species. Native species are accustomed to transpiring only during the rainy season. Indigenous species normally loose their leaves or become dormant during periods of low flows (Sandström, 1995). If the rainy season, where the exogenous species comes from, is out of phase with the rainy season of their current environment the tree transpires during the dry season, reducing the base flow. Additionally, it has been observed that the riparian vegetation induces a diurnal fluctuation on the base flow, related to the diurnal transpiration rates of the trees in the riparian zone. Several studies already indicate that only a small percentage of the total area is responsible for these fluctuations in the stream flow (Bond et al., 2002; Bren, 1997; Dye and Poulter, 1995), which can amount up to 50 percent of the base flow.

Changes in runoff responses in a catchment are often linked to land use changes, however, it is not only land use, which affects the runoff behaviour of a catchment. Changes in land use are often related to changes in vegetation cover, changing the first rainfall partitioning point by altering the interception. In addition, the second rainfall partitioning point is altered by the change in transpiration. Land use change changes the soil properties such as bulk density and saturated hydraulic conductivity (e.g. Bronson et al., 2004; Franzluebbers et al., 2000; Murty et al., 2002; Sandström, 1995). Bormann et al. (2007) showed that these second order changes contribute significantly to altering runoff responses. In addition, Fenicia, et al. (2008) argued that the age of a forest stand may also have an impact on the runoff response (young trees transpire more than old trees). Lørup et al. (1998) showed that for a catchment in Zimbabwe, without land use change, the runoff response changed significantly, which they attributed to the change in agricultural land management. Different soil and water management techniques applied in agriculture have demonstrated a change in rainfall partitioning (Kosgei et al., 2007; Makurira et al., 2007b) and therefore in runoff generation. Sandström (1995) showed in a catchment that the responses from a deforested catchment are a result of a combination of changes in the two rainfall partitioning points. Depending on these changes, which can have an inverse relationship, the runoff response changes accordingly (Brown et al., 2005). Sandström (1995) showed in the Babati district in Tanzania, that the deforested catchment generates the same amount of runoff as the forested catchment. He suggested that initially more water is generated in the catchment directly after deforestation. Infiltration rates are still high and the transpiration has reduced significantly. However, in his research catchment, the soils are slowly degraded through cattle grazing, decreasing the infiltration rates and compacting the soil. Rainfall either runs off quickly or remains in ponds and water in the topsoil evaporates within a couple of days. At the final degraded stage, the total annual runoff, in this case, is in the same range as the runoff generated in the forested catchment, only the temporal distribution of the flows is different. The degraded deforested catchment is characterised by quick and high peak flows, whereas the forested catchment retains the water some time before it is drained from the groundwater storage. For both the deforested-degraded and forested catchment the annual evaporation (combination of interception, soil evaporation and transpiration, see Equation 1.2) is the same.

Finally, although most studies show that population growth and corresponding changes in land use are mostly associated with negative impacts. A study in Machakos, Kenya, (Tiffen et al., 1994) describes that an increase in population instead of leading to environmental degradation can lead to environmental recovery. Increasing population density initially increases environmental degradation; people have the option to move to new more suitable areas. The process of environmental recovery is set in motion when the population density reaches a certain level; people are forced to come up with new technologies to prevent degradation and reduction of land productivity. Adoption of new technologies is improved by migration, education, population density (increased labour force) and a market for agricultural products, which are all enhanced by a population increase. A high population density allows the people to interact more frequently and therefore enhancing the transition of knowledge. With the right incentive (reduction of land productivity and degradation) and the right environment (access to information and money to invest) people can fight environmental degradation.

1.3 HYDROLOGICAL PROCESS UNDERSTANDING

In many parts of the world, catchments are ungauged and this is particularly true for catchments in the developing world, where gauging networks are in decline as a result of lack of financial and human resources (Mazvimavi, 2003). This severely hampers runoff prediction in these catchments. One of the ways to improve runoff prediction in these areas is through improved understanding of the hydrological processes (Sivapalan et al., 2003). The research catchment was ungauged until the beginning of the current SSI programme (Rockström et al., 2004), there were only three rainfall stations in and near the catchment and no discharge stations. Analysis of land use changes and the impact on runoff prediction in this research catchment is difficult without the existence of long term discharge data. Understanding of the hydrological processes in the area required additional information beyond conventional data such as rainfall and discharge data, which are still invaluable. Hence the SSI programme installed at several locations discharge measuring stations (Bhatt et al., 2006). In addition, a multi-method approach (Blume et al., 2008) has been used to understand the dominant hydrological processes in this catchment. This method can be used in ungauged catchments to understand the runoff generating processes and predict runoff.

To investigate the dominating hydrological processes and rainfall runoff relationships, different types of field studies have been conducted. It started with setting up the conventional rainfall and runoff network; (1) regular rainfall and runoff measurements at a short temporal and spatial resolution during the programme time span (Bhatt et al., 2006); (1a) Rainfall was measured at daily time steps in upto 40 locations of the catchment, hourly rainfall was measured at 6 locations in the catchment (Chapter 3). (1b) An automatic meteorological station recorded hourly data on meteorological parameters to determine Penman-Monteith's evaporation (Chapter 3). (2) A nested catchment was established (Brown et al., 1999; Didszun and Uhlenbrook, 2008; Jewitt and Görgens, 2000; McGlynn et al., 2003; Shanley et al., 2002), in which reliable data have been collected over a period of 2 years (some data was lost due to extreme flood events (Mul et al., 2008b Chapter 4)). High resolution data collection (hourly rainfall and runoff data) started early 2006. The smallest size catchment was conjunctively used for a paired catchment study.

(3) A multi-method approach has been adopted to develop the hydrological understanding of the system through several concentrated field work activities. The activities focussed on four different aspects; (3a) The groundwater flow system has been studied through geological mapping, spring water sampling and geophysical investigations (Mul et al., 2007a Chapter 4). (3b) Several single events were investigated, two relatively small floods were analysed for pre-event and event water (Mul et al., 2008a Chapter 5), and an extreme event was investigated for spatial rainfall variability, peak flow reconstruction, and hydrograph separation (Mul et al., 2008b Chapter 5). (3c) Base flow fluctuations were observed for the smallest size catchment, and investigated (Mul et al., 2007b Chapter 4). (3d) The understanding of the hydrological processes were then integrated into a conceptual hydrological model (Mul et al., 2008c Chapter 6).

Each experimental method has its own strengths and shortcomings concerning costs and the temporal and spatial scale at which they can be used. Until now, these methods have been mainly developed and applied in humid temperate climates. In arid and semi-arid areas, two additional factors make the process investigation difficult: First, the climatic spatio-temporal variability is very high and, second, human influence often due to increasing population densities results in rapid land use and stream channel changes that affect hydrological responses.

In the research area high population pressure not only changes the land use and management, as described in section 1.2, it also affects the stream flow in a more direct way. Storage structures have been constructed to store water for domestic use, which is then redirected to communal taps. However, the water used for these systems is small in comparison to the amount which is used for agriculture. In the area, historically, the agricultural lands have been irrigated, through full or supplementary irrigation. Rainfall and river flow variations dictate the pre-dominant irrigation type in a specific area (Chapter 7).

The improved understanding of the hydrological processes will be beneficial to predict the impact of different smallholder system innovations on the overall water balance. Additionally the improved understanding of the hydrological processes will improve the hydrological modelling and confirm the hydrological conceptualisation, which is required to understand the implications of a successful uptake of the selected innovations. The different methods used and the results from the analyses can be seen in Chapter 4, 5 and 6.

1.4 Hydrological Modelling

Hydrological modelling is applied to verify the hydrological process understanding developed during the study period. Several problems arise when working with hydrological models, such as data availability (especially when working with physically based models), spatial diversity between observed parameters and model parameters, differences between hydrological process scales and modelling scales. The choice of a model is determined by the purpose of the model and the availability of data.

Different spatial and temporal scales have serious implications on hydrological models. Conceptual models applied to different size catchments can be very different although all of them model the respective size accurately. However, when transferring one modelling outline to another scale it may be difficult to obtain accurate output. Jothiyangkoon et al. (2001) showed that a conceptual model including saturation overland flow and evaporation is adequate for predicting inter-annual variability. However, when looking for intra-annual variability the model does not perform equally well. Adding the processes of sub-surface runoff, and separating evaporation and transpiration improves the model at this scale. Again for a model to calculate daily timesteps additional processes need to be added. A non-linear storage-discharge relation of the subsurface flow and deep groundwater storage is needed to capture the daily variation. The research by Jothiyangkoon et al. (2001) only looked at different timescales of hydrological responses, however, the spatial scale is very much linked to the temporal scale through the residence times in the catchments. The response time of a catchment is, next to catchment characteristics, such as hill-slope response, channel hydraulic response and network geomorphology, directly related to the size of a catchment. In this regard, responses from small catchments are related to within storm patterns of rainfall. At a larger scale catchment seasonality becomes more important.

These differences of scales for the hydrological processes influence hydrological modelling. Large-scale hydrological models tend to need less complexity for correctly modelling the hydrology than small-scale hydrological models. Several authors (Jewitt and Görgens, 2000; Savenije, 2001; Sivapalan, 2003) recognised that when moving to a larger scale, hydrological processes appear to become more regular than the processes at a small scale. At a small scale variability in the catchment characteristics has a big influence on the hydrological response. At large scales this variability is averaged out. As a result the parameters obtained can not be determined by local field observations. The variability is already incorporated in these parameters. Two approaches exist to link the hillslope processes with the large scale observed processes.

The first approach is the downward approach (Klemeš, 1983), where the large-scale response is derived from the observed hydrological response of a catchment. The hydrological model representing this scale is a conceptual model only incorporating basic rainfall-runoff relations. These large-scale hydrological models are relatively easy to set-up as the necessary data is in most cases available. This model is then tested at a smaller scale where necessary additional processes are added. Jothityangkoon et al. (2001) showed a good example of this type of approach.

The second approach is the upward approach (Reggiani et al., 1998; Uhlenbrook et al., 2004), where the small-scale hydrological model is based upon the observed hillslope hydrological processes. The difficulty is that only few hillslopes have been intensively monitored and with a large variety between catchments, it is difficult to model the hydrological response accurately. When upscaling this concept to larger scales, it becomes apparent that additional processes such as channel processes need to be taken into account (Uhlenbrook et al., 2001). Other tributaries joining the main river not necessarily have the same observation network and may cause additional errors.

Both approaches suit the initial purpose, in the downward approach, to predict the hydrological response of the watershed and in the upward approach to improve the knowledge on hillslope processes. Sivapalan (2003) identified that, although the knowledge of hillslope processes is increasing, the link between these processes and the hydrological response on watershed level is not determined. He indicated that there is quite a large gap in this area of research. He suggests that in order to get a better understanding of the linkages between the response on watershed level and the hillslope processes, the two approaches should be combined. Modelling the hydrological response at large scale through conceptual models based upon observed rainfall runoff relation and modelling the hydrological response at small scale through detailed modelling of hillslope processes.

In this study, the hydrological process understanding generated by the multi-method approach has been used for the development of a conceptual hydrological model at the mesoscale. The approach followed in this research is the downward approach (Fenicia et al., 2008; Jothityangkoon et al., 2001; Klemeš, 1983), where step by step complexity is added. A basic conceptual model with two buckets, one for the unsaturated zone and one for the saturated zone, has been developed first (Fenicia et al., 2007; Winsemius et al., 2006). Discrepancies between observed data and the model were analysed, after which the model structure was altered to incorporate processes observed in the catchment. One of the main advantages of using the downward approach is that the amount of modelling parameter is limited, and only complexity is added when it is required. Constraining the model for several indicators can reduce the equifinality of the model (Beven, 1993; Beven and Binley, 1992; Kuczera and Mroczkowski, 1998; Mroczkowski et al., 1997; Savenije, 2001), and therefore increased prediction efficiency (Uhlenbrook and Sieber, 2005; Winsemius et al., 2006). Observed hydrological processes can contribute in reducing the equifinality.

1.5   OBJECTIVES

Three main objectives are addresses in this study, which attempts to understand the linkages between the smallholder farmers living in the area with the environment they live in. This linkage is a feed back linkage, where climate and local condition dictate the farmer's activities and requirements. At the same time, farmer's activities impact the resources availability, which changes the conditions for other farmers, who in turn adjust their activities accordingly. As can be seen from the above, this is a dynamic process, which has many driving forces from outside. As water is the constraining factor, the hydrological linkages are very important. To understand these hydrological linkages, the hydrological processes need to be understood.

This thesis highlights the hydrological aspects of the farmer's options, develop hydrological process understanding of the catchment and give a qualitative assessment of the impact of the farmer's activities. The assessments are made during a period of four years in a previously ungauged catchment. The following objectives have been identified:

1) This study investigates the constraints and opportunities that smallholder farmers face in the study area as a result of spatial heterogeneity in the catchment, both climatological and biophysical constraints.

2) This study investigates the hydrological processes governing the occurrences and the temporal variability of the water availability.

3) This study investigates the impacts of farming activities on the hydrology at catchment scale.

Chapter 2

STUDY AREA

The study area is part of the Pangani river basin, many of the issues in the study area are applicable to the entire Pangani basin, which is highlighted in this chapter.

2.1   PANGANI RIVER BASIN

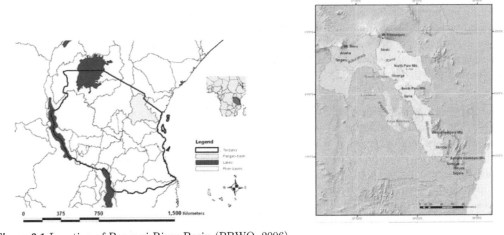

Figure 2.1 Location of Pangani River Basin (PBWO, 2006).

The study has been conducted in the Pangani river basin, one of nine major river basins in the country, located in northern Tanzania. The Pangani (or Ruvu) River has an estimated catchment area of 42,200 km^2, with the major part located in Tanzania and a small part in Kenya (Fig. 2.1). The basin can be sub-divided into 5 main catchments and 14 sub-catchments (PBWO, 2006). The main catchments (Ruvu, Kikuletwa, Pangani main stream, Mkomazi and Luengera) have distinct hydrological characteristics. The Ruvu and Kikuletwa catchments originate from the two highest mountains in Tanzania (Mt. Meru and Mt. Kilimanjaro). These catchments are characterised by many perennial springs, which are fed from the mountains, and in the case of Mt. Kilimanjaro, snow and ice melt in the upper parts, is an important component for the runoff generation. These catchments flow into the Nyumba ya Mungu dam, which was mainly constructed for hydropower generation. The live storage capacity of the Nyumba ya Mungu Dam of 1,100 million m^3 is equivalent to about 71 percent of the mean annual runoff (PBWO, 2006). Currently this dam generates about 17 percent of the country's hydropower (IUCN, 2003). Downstream of the dam is a large wetland which used to periodically inundate during the wet season. However nowadays, the flooding is determined by the dam releases, which tries to optimise the releases to supply the downstream hydropower plant (New Pangani Falls). After Nyumba ya Mungu dam there are only two major tributaries which contribute substantially to the Pangani River. These are the Mkomazi and Luengera catchments originating in the South Pare and Usambara mountains. Some local systems contribute along the main stream. This comprises of the dry plains on the west side of the river and some localised flow systems originating from the western side of the Pare Mountains. Generally, the streams coming from the plains are intermittent and only produce runoff during the wet season. The streams coming from the western side of the Pare Mountains recharge the local groundwater and wetland, and do not have a considerable surface water contribution to the Pangani River.

The average annual rainfall has a large variation through the river basin, ranging from 400 mm a^{-1} in the semi-arid area of the steppe in the Pangani Main stem catchment to nearly 2,000 mm a^{-1} on the slopes of Meru and Kilimanjaro (Norbert et al., 2002). The rainfall pattern in many parts of the basin is bi-modal with a short and long rainy season. The short rains occur mainly between October and January, locally known as '*Vuli*', and the long rains occur between February and May, locally known as '*Masika*'. Rainfall patterns in the Pangani River Basin have a high spatial and temporal variability. Annual rainfall variability can be described roughly using altitude and distance to the ocean (Fischer, 2008). Daily rainfall is much more localised, particularly during high intensity storms, whereby altitude is not necessarily linked to the amount of rainfall (Mul et al., 2008b). Additionally, the temporal variability is high, sometimes with standard deviation equal to the average rainfall. Dry and extremely wet years are common, having disastrous effects on the local population; droughts causing moisture stress and floods causing water logging, soil erosion and structural damage; both eventually leading to crop failure.

The Pangani basin has an estimated population of 3.7 million people (IUCN, 2003) of which 90 percent lives in rural areas, although the urban population in the two major cities (Arusha and Moshi) is rapidly growing. Eighty percent of the rural population depends, directly or indirectly, on agriculture for their livelihood (Mwamfupe, 2002). Traditionally, people only cultivate crops in the high areas, where conditions were favourable (higher rainfall, lower temperatures and no malaria). Additionally, it is said that lions did not live in mountainous areas. The cattle rearing Masai occupies the plains. With the increase of population in the higher areas, people were forced to move to the lower areas and start permanent cultivation in the less suitable areas. Both in the higher and lower areas traditional irrigation systems have developed, many of the irrigation canals and intakes are indigenous, with the intake structures made of logs, mud and stones and the canals being mostly unlined earthen furrows (Adams et al., 1994; Fleuret, 1985; Mul et al., submitted). Intake structures are regularly washed away and must be rebuilt every year. The efficiency of these systems is very poor, with measured water losses as high as 80 to 90 percent (IUCN, 2003; Makurira et al., 2007a; Turpie et al., 2003). However, data on water losses always needs careful interpretation because of the different ways water losses can be and have been defined. In recent decades, several NGOs and governmental organisations such as JICA and SAIPRO have assisted water user groups (WUGs) in rehabilitating storage structures, strengthening intakes and lining canals. These improvements aim to increase system efficiencies. However, an improved efficiency of an irrigation system does not automatically lead to a reduction of the amount of water abstracted, but often rather to an expansion of the command area. An unintended consequence of the modernization of furrow irrigation may then be that equity decreases amongst WUGs because upstream WUGs are now able to abstract more water from the river while return flows are reduced, depriving downstream users. This has been documented for the Rufiji river basin (Mehari et al., 2008).

Many of the several thousand irrigation canals in the Kilimanjaro region were built before the onset of the 20th century, while irrigation itself is traced back several hundred years. An inventory of the irrigation furrows in Kilimanjaro Region in 1977 showed more than 500 indigenous irrigation furrows with approximately 1,800 km of main canal abstracting up to $200*10^6$ m^3 a^{-1}. In 2003, more than 2,000 indigenous irrigation furrows were found in the higher areas of the Pangani River basin (Turpie et al., 2003). Most are found on the slopes of Mt Kilimanjaro and Mt Meru, but many others can also be found in the Pare and Usambara Mountains. Moreover, indigenous smallholder irrigation accounts for 90 percent of water consumption by irrigation users, with 85 percent of water consumption in the Kilimanjaro region attributable to irrigation (URT, 1977). These water abstractions and climate change over the past decades have reduced instream flows into the Nyumba ya Mungu reservoir from several hundred to less than 40 m^3 s^{-1} (IUCN, 2003), affecting the reliability of power supply. Competition over water between the irrigation systems is increasing as the irrigation areas in the upper reaches of the Pangani Basin keep on expanding (Grove, 1993; Potkanski and Adams, 1998).

Not only does the increasing population and increasing food requirement have a direct impact on the water resources through abstractions, it also has an impact on the land use in the basin. The forest cover in the basin is increasingly threatened by encroaching agriculture. Between 1952 and 1982, Kilimanjaro's natural forest area declined by 41 km^2 (IUCN, 2003). Assuming that the Pare and Usambara Mountains were, prehistorically, more or less continuously covered by forest, 77 percent of this has been lost to human disturbance and fire over the past 2,000 years (IUCN, 2003). The Pare and Usambara Mountains occupy some of the most densely populated areas of the Pangani basin (IUCN, 2003). On the West Usambara Mountains, populations have grown 23-fold since 1900 (Newmark, 1998). The pressure continues on the basin's remaining forest reserves.

2.2   SOUTH PARE MOUNTAINS

The research focuses on the South Pare Mountains in the mid-reaches of the Pangani basin. Existing information on geology, rainfall, runoff and water users was gathered and analysed to create and overview of this area. The most important features are the geology and rainfall occurrences, which dictate the water availability and requirements.

2.2.1 Geology[1]

The geology of the South Pare Mountains has a big impact on occurrence of streams and runoff generation (Mul et al., 2007a). It is mainly composed of folded and weathered metamorphic igneous rocks and superficial deposits (Bagnall, 1963). The metamorphic rocks are characterised by meta-sedimentary and meta-igneous types of rocks that are associated with the granulite-gneiss complexes in the Mozambique belt (Bagnall, 1963; Muhongo and Lenoir, 1994; Mutakyahwa et al., 2001). Rocks have undergone regional metamorphism resulting in the development of a strong foliation almost parallel to the bedding with a predominant granulite-pyroxenes facies. The geology is dominated by a large recumbent fault oriented towards the northeast with an angle of 25 degrees. In addition there are minor faults facing the same direction with angles of more than 40 degrees (URT, 1965). The steep scarps observed in the catchment are retreated fault scarps, with the actual faults buried under superficial deposits. The soils on the forested summit area and the scarp slopes of the South Pare Mountains are red loamy clays; surface limestones occur sporadically. A series of NW-SE trending faults cuts the study area, retaining the initial strike direction of the major fault (Bagnall, 1963; Muhongo and Lenoir, 1994).

The main dipping direction of the rock layers is between $25°$ and $50°$ E. Faults that intersect the area largely define the drainage network of the area (Fig. 2.2). Soils on the forested hill summits and the low-lying areas of the South Pare Mountains are characterised by dark-reddish to reddish-brown colour. In contrast, soils of the scarp slopes and some of the lower parts, are predominantly yellowish. In the lowlands, unconsolidated deposits are found, consisting of reddish soils. In the dry river beds, alluvial deposits are found, which are a result of rapid, violent erosion and deposition (Fig. 2.3).

---

[1] This section is based on: Mul, M.L., Mutiibwa, R.K., Foppen, J.W.A., Uhlenbrook, S., Savenije, H.H.G., 2007a. Identification of groundwater flow systems using geological mapping and chemical spring analysis in South Pare Mountains, Tanzania. Physics and Chemistry of the Earth 32 (15-18), pages 1015-1022.

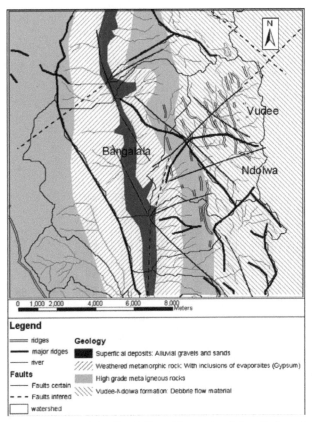

Figure 2.2 Geological map of a part of the South Pare Mountains (after Mul et al., 2007a).

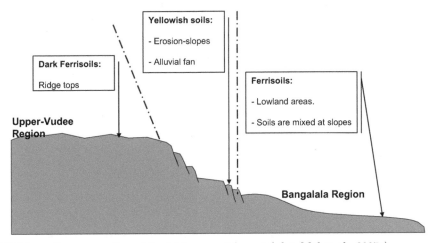

Figure 2.3 Schematic cross section of the Makanya catchment (after Mul et al., 2007a).

2.2.2   Rainfall

Average monthly rainfall and standard deviations are depicted in Fig. 2.4 for the Same meteorological station, located at the foot of the South Pare Mountains. Although it is not apparent in the average monthly rainfall series, generally there is a short period between the two seasons, *Vuli* and *Masika,* which is dry (January and February).

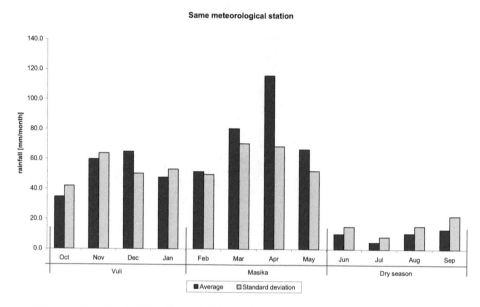

Figure 2.4 Distribution of rainfall in Same (1934-2007).

Historical rainfall data (1934-2007) has been used from Same meteorological station (altitude 882 m). Two other rainfall stations within the South Pare Mountains have also been used, which have records from the early 1990s – Tae Malindi (altitude 1,740m) and the sisal estate in Makanya (altitude 700m). A correlation analysis between the three stations shows a good correlation (between 0.70-0.75 on a monthly basis, for the period 1935-2006), although the absolute values differ substantially, Tae (796 +/- 228 mm a^{-1}), Same (556 +/- 183 mm a^{-1}), Sisal Estate (521 +/- 285 mm a^{-1}). The long term daily data from the Same rainfall station (assumed to be representative of the South Pare Mountains) was used for trend analysis.

Table 2.1 Average rainfall and standard deviation of the rainfall, Same station.

| | Annual (mm a^{-1}) | | Vuli (mm season^{-1}) | | Masika (mm season^{-1}) | |
	Average	St Dev	Average	St Dev	Average	St Dev
1935-1940[2]	599		127		429	
1940s[3]	532	166	181	121	317	144
1950s	555	137	179	76	358	133
1960s	603	188	247	155	324	181
1970s	615	250	241	162	339	135
1980s	587	140	216	90	343	105
1990s	499	276	213	240	264	66
2000-2007[4]	491		225		238	
1935-2007	561	184	208	141	327	131

The meteorological data (Table 2.1) suggest that rainfall in the area increased between the 1930s and 40s, with highest 10-year averages in the 1960s and 70s, reducing after 1980, with the lowest of 491 mm a^{-1} in recent years (Table 2.1). Spearman rank tests were performed on annual totals, *Vuli* and *Masika* seasons. The test for annual totals showed a visible but not significant trend (Spearman rank test, $t = -1.54$ within the interval -2 and 2 (2.5 percent significance level)). Fig. 2.5 showing a plot of annual and seasonal rainfall (1934-2007) suggests that since 1990, a period with clearly below average rainfall dominates, with the exception of the 1997-1998 El Niño season. This confirms the general opinion of the farmers in the area (Enfors and Gordon, 2007). The tests for *Vuli* and *Masika* also show no significant trend ($t = 1.18$ and -1.94, respectively), although Fig. 2.5 indicates a visible declining trend for the *Masika* season, whereas *Vuli* rainfall shows an increasing trend. The variations in *Vuli* are higher, with an average rainfall of 210 mm season^{-1} and standard deviation of 140 mm season^{-1}, than in *Masika*, with a higher average rainfall of 330 mm season^{-1} and a lower standard deviation of 130 mm season^{-1}.

---

[2] 5 years of data
[3] Decade starts October 1940 until September 1950
[4] 6 years of data

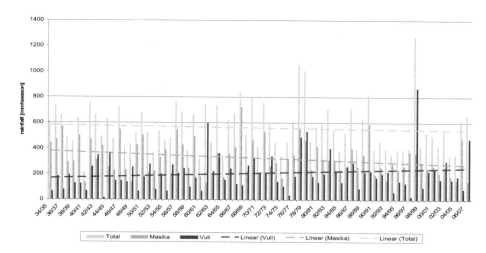

Figure 2.5 Annual, *Masika* and *Vuli* rainfall at Same station showing trend lines.

Secondly, a trend analysis has been done on the daily maximum rainfall for annual, *Masika* and *Vuli* values (Fig 2.6). The annual and *Masika* daily maximum rainfall show a visibly declining trend, whereas the *Vuli* daily maximum does not show any trend. However, the Spearman rank tests do not indicate significant trends (annual $t = -1.62$, *Masika* $t = -1.61$, *Vuli* $t = 0.16$, all within the range of -2 and 2).

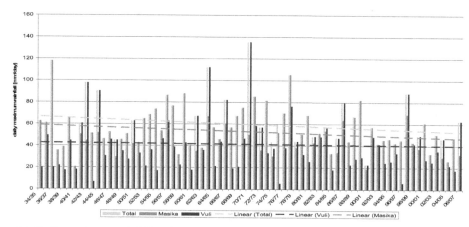

Figure 2.6 Daily maximum rainfall at Same meteorological station.

Finally, a trend analysis has been done on the number of rainy days. It shows that the number of rainy days in *Vuli* are increasing and in *Masika* these have remained stable (Fig 2.7). On annual basis the number of rainy days increases as well. The Spearman rank test has also been performed on these data series which shows that for *Vuli* season the trend is significant, with $t = 2.47$. The trend in both annual and *Masika* season are not significant, $t = 1.47$ for annual and $t = 0.07$ for *Masika* season.

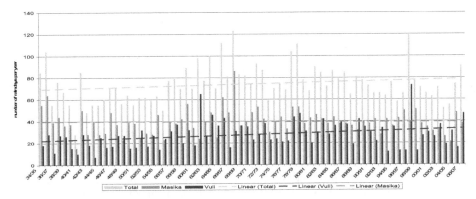

Figure 2.7 Amount of rain days per year at Same meteorological station.

Although the above analysis only shows a significant increasing trend in the number of rainy days for the *Vuli* season, several of the Spearman rank tests show a visible trend (although not significant). Annual and *Masika* rainfall totals show a decreasing trend, similar to the annual daily maximum rainfall.

This analysis is consistent with the study performed by Valimba (2004) and Enfors and Gordon (2007) indicating that in northern Tanzania changes in rainfall patterns are not significant for annual totals. However, they found that the occurrences of low intensity rainfall (<10 mm day^{-1}) have significantly decreased, whereas high intensity rainfall has increased. The analysis for the Same meteorological station (Table 2.2) shows that there is a variation in the number of days with high rainfall events during the period of observation, but there is no trend. This difference is probably due to the inclusion of the 1940s and 1950s in the analysis (this data was received from PBWO). Both 1960s and 1970s were relatively wet, a trend analysis starting from the late 1950s shows therefore a different picture.

Table 2.2 Rainfall occurrences for Same meteorological station.

	n >0.05	n >1	n >10	n >20	n >30	n >40	n >50	n >60	n >70	n >80	n >90	n >100
1940s	591	476	172	81	40	21	10	5	3	3	2	0
1950s	652	487	162	76	44	21	14	11	4	1	0	0
1960s	900	596	165	80	38	21	12	8	3	3	1	1
1970s	825	573	174	86	50	29	20	13	6	4	2	2
1980s	794	553	171	77	35	20	10	6	1	0	0	0
1990s	742	539	162	67	37	25	8	6	6	2	0	0

Enfors and Gordon (2007) observed for Same (1957-2004), that changes in the rainfall pattern are mainly apparent in the larger occurrence of long dry spells (>21 days) during the *Masika* season. Long dry spells have increased in occurrence significantly from 20 percent before the 1980 to 80 percent after 1980. These long dry spells are disastrous for farmers, whereby only supplementary irrigation from groundwater, surface water or rainwater harvesting structure, can prevent complete crop failure.

2.2.3   Agricultural water use

In general, yellow soils, which cover a large part of the valley area, have poor nutrient composition and water holding capacity (Enfors and Gordon, 2007). Near the river, alluvial deposits have much higher fertility and water holding capacity. Due to these soil types and rainfall variability, the probability of crop failure is high in the Pare and Usambara Mountains (Norbert et al., 2002). Thus, the challenges facing the people in the South Pare Mountains are very similar to the problems in other parts of the Pangani basin. Population growth and the resultant need for more agricultural land has put the remaining forest cover under immense pressure. Already the loss of natural forest cover in the South Pare Mountains is estimated to be as high as 73 percent (Newmark, 1998). In addition, farm plot sizes are reducing, increasing the need for high productivity per unit of area. The practice of rotational cropping is also under threat, leaving the soils no time to recover, resulting in depleted soils and increased exposure to soil erosion. There is a need for supplementary irrigation or rainwater harvesting techniques to produce a reliable crop growth since rainfall in general is not sufficient for cultivating maize, the preferred crop (Makurira et al., 2007b).

Because of the bi-modal rainfall pattern, the people in the South Pare Mountains grow a short period crop twice a year. Supplementary irrigation through indigenous furrows has been practiced in this area for more than one hundred years. Some of the furrow systems include a storage structure, which provides extra hydraulic head to supply the farthest downstream farmers in the system (Makurira et al., 2007a). NGO's have been instrumental in enlarging these storage structures and lining canals (TIP, 2004). Three areas can be identified where irrigation is applied in different periods; highlands (1,200m and above), midlands (in the valley) and lowlands (area around Makanya village).

In the highlands, agriculture is practiced throughout the year, with indigenous furrows diverting water for supplementary irrigation during the two rainy seasons and full irrigation during the dry season. From a steep escarpment these perennial rivers flow into the midlands where the water is used for supplementary irrigation during the rainy seasons. In general, the command area and the capacity of the irrigation system are out of sync. Ideally during a dry spell all the farms are irrigated, however one system in Bangalala shows that less than 10 percent of the plots are irrigated when it is most needed (Makurira et al., 2007a). Additionally, during these dry spells runoff from the rivers is low, and even less water is available. Competition for water during these periods is extremely high (Makurira et al., 2007a). The remaining water, usually the leakages from the diversion structures, continues its flow downhill until it reaches the valley of the catchment where the majority of the runoff recharges the local aquifer under the sandy river bed (Mul et al., 2007a). Only flood flows reach the outlet of the catchment which the lowland farmers divert into their plots for irrigating crops such as maize and beans. This type of irrigation is known in literature as spate-irrigation (see Mehari et al., 2005).

2.2.4   Runoff

High intensity rainfall events can cause flooding in the downstream. According to PBWO (2006), the South Pare Mountains is the only area in the Pangani River Basin which experiences large flood events even during the short rainy season as observed by Mul et al. (2008b) for the Makanya catchment. Rivers originating in the South Pare Mountains flow in two main directions, east and west. Different river systems join at the eastern as well as the western side of the South Pare Mountains and flow to the South towards the main Pangani River. A large amount of runoff generated in the mountainous area does not appear at the outlet of the catchment. In addition, water using activities in the area reduce the flows to the mid- and lowlands. This runoff is captured before reaching the Pangani River through riparian infiltration and periodic flooding of non-draining swamp area (PBWO, 2006).

The east side of the South Pare Mountains drains into Mkomazi River, which is the largest river contributing to the Pangani River downstream of Nyumba ya Mungu Dam (PBWO, 2006). Highland marshes located on top of the Shengena Mountains predominantly drain to the east of the South Pare Mountains, and thereby provide a continuous supply of water. Also in this catchment, there has been an increase in water use (58% of the runoff generated is used) resulting in reduced seasonal flows into the floodplains and thus changing perennial flows into the Pangani (PBWO, 2006). Fig. 2.8 shows the difference in the naturalised flows, without water users in the catchment compared to the observed flows. It can be seen that not only the flows are reduced, during the dry season the river now often dries up.

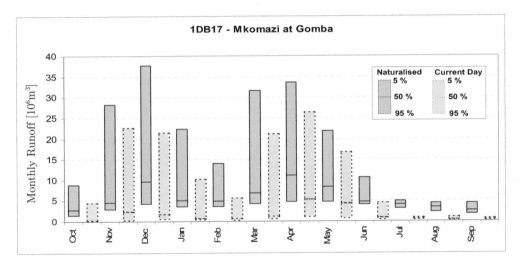

Figure 2.8 Comparison of Naturalised and Current Day Flows – Mkomazi catchment (PBWO, 2006).

The western part of the South Pare Mountains drains into the Makanya River, part of the Pangani Main stem catchment area. Four perennial rivers drain the mountainous area - Mwembe, Vudee, Chome and Tae River. In turn, they join in the valley of the catchment, where most of the runoff generated in the highlands recharges underlying aquifers contained in the alluvial deposits along the river valley. When there is substantial runoff, water may reach Makanya village positioned at the outlet of the catchment, where farmers apply spate-irrigation techniques to utilise the water in a short period (usually flows only last several days). SUA (2003) reported that even if the water reaches Makanya Village, it rarely reaches the Pangani River, which is 40 km downstream. The water disperses and infiltrates into the sandy soils, once characterised as seasonal wetlands. Historically, the runoff from the mountain used to reach Makanya village throughout the year (Mul et al., 2006). Population increase and resultant water use increase have triggered these changes.

The population in the Makanya catchment is estimated to be 35,000 and is rapidly growing with an estimated growth rate of 1.6 percent per annum (URT, 2004). Increasing population density has increased the necessity for agricultural lands. In the upland areas, steep slopes have been cleared of trees and are subsequently cultivated (Enfors and Gordon, 2007). Seasonal rainfall in general is not enough for growing the most popular crop, maize, which emphasises the necessity for full or supplementary irrigation (Makurira et al., 2007a). The abstractions in the upland and midland have reduced base flow significantly (Mul et al., 2006). There are over 100 indigenous furrows in the in Makanya catchment, each supplying water to areas ranging from 0.5 to 40 hectares. Similar to the Chagga systems described by Grove (1993), people use furrow water not only for irrigation but also for domestic use and for watering livestock. Many irrigation furrows also have micro dams (75 have been identified in Makanya). These micro dams are mostly located in the upstream parts of the command area of a furrow system and serve to temporarily store water when nobody irrigates, in order to boost the diverted river flow in the furrow when farmers are irrigating. Without such reservoirs the water would not reach the most distant users because of the large transmission losses (Makurira et al., 2007a).

Chapter 3

WEATHER AND CLIMATE

Hydrological measurements have primarily been done in the western catchment of the South Pare Mountains, the Makanya catchment. This consisted of meteorological and discharge measurements (the latter is elaborated in Chapter 6).

3.1   RAINFALL

The long term rainfall measurements in Same, Tae and the Sisal estate are insufficient to capture the spatial rainfall variability in the catchment. With the SSI-research focus located in Bangalala and Mwembe, in the upper and middle parts of the Makanya catchment, additional rain gauges were necessary. In 2004, the SSI programme installed additional manual rainfall stations (see Fig. 3.1 and Table 3.1). The manual rainfall stations are read every day at 9 AM, similar to the official meteorological stations in Tanzania. An automatic meteorological station was installed in Bangalala, measuring a range of meteorological data such as humidity, temperature, wind speed, solar radiation and rainfall intensity. In February 2006, 5 more automatic rain gauges were installed in the catchment, measuring rainfall intensity (see Table 3.1).

Table 3.1 Rainfall stations in Makanya catchment

Rainfall station	Altitude
	m
Tae Malindi	1741
Chome	1664
Ndolwa (+ automatic)	1544
Heri Vudee (+ automatic)	1503
Tae primary	1417
Vudee (+ automatic)	1396
Chani	1306
Mwembe (+ automatic)	975
Iddi	960
Bangalala (+ automatic meteorological station)	938
Mchikatu (+automatic)	885
Same	882
Eliza	870
Wilson chini	834
Sisal estate	698
Makanya	640

Figure 3.1 Rainfall stations in Makanya catchment installed in 2004.

Fig. 3.2 shows the monthly and cumulative rainfall of the respective stations, the bars with the reddish colour represent the mountain stations (Tae Malindi, Chome, Ndolwa, Vudee and Chani) the green bars represent the midland stations (Mwembe, Iddi, Bangalala, Mchikatu, Eliza and Wilson chini) and the blue ones represent the stations that are located in the lowlands (Same, Makanya and Sisal Estate). The station Tae primary school was taken out of the analysis because of inconsistencies with surrounding stations[5]. Chome and Ndolwa are not depicted in Fig. 3.2a because they were installed in 2005.

As a general trend, it can be seen that mountain stations on a monthly basis record more rainfall. The three stations located in the plains record the lowest rainfall. However, during individual rain events, particularly rainfall with very high intensities, this is not always the case (Mul et al., 2008b). The graphs also clearly show the large variability between the different years and seasons: *Vuli* (October-January) in 2005 (Fig. 3.2b) was clearly a below average season, whereas *Vuli* 2006 (Fig. 3.2c) was extremely wet (coinciding with an El Niño year). During these four years of observation only 2006 had good rains, the others were regarded as dry years. Although total seasonal amounts are important for food production, even more important are the inter-seasonal variations, particularly the occurrence of dry spells, which can not be seen in these graphs.

---

[5] e.g. during an exceptional event it recorded 22 mm d^{-1}, whereas the surrounding stations Tae Malindi recorded 106 mm d^{-1} and the lower station Sisal estate 56 mm d^{-1}, additionally it recorded nearly 700 mm month^{-1} in November 2006, which was more than twice any other station recorded.

a

2004/05

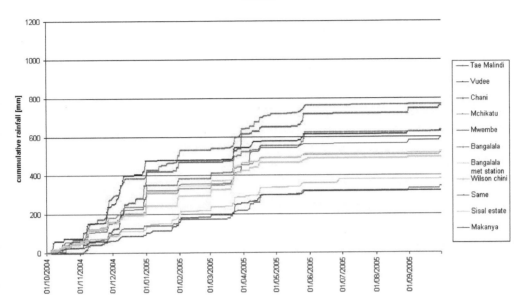

monthly rainfall 2004-05

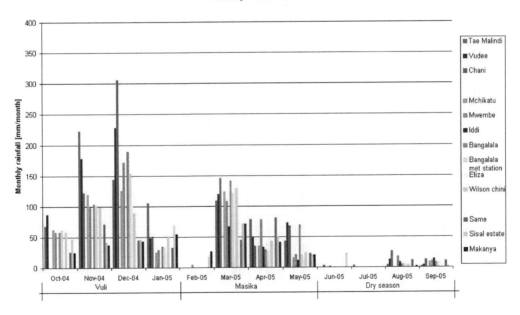

b

2005/06

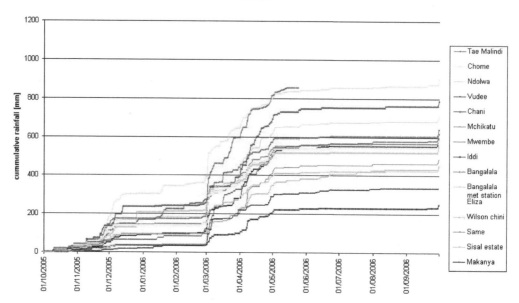

monthly rainfall 2005-06

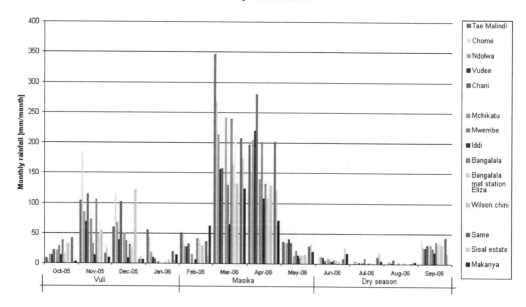

c

2006/07

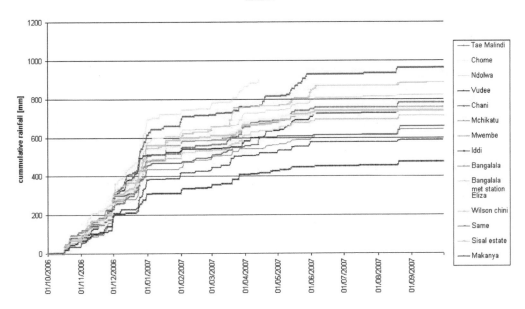

monthly rainfall 2006-07

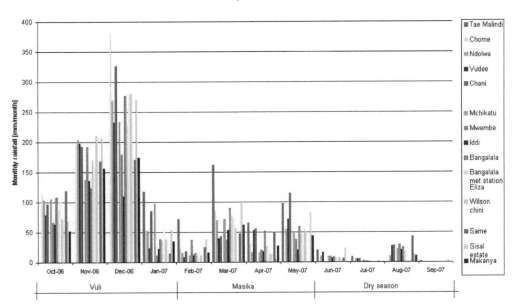

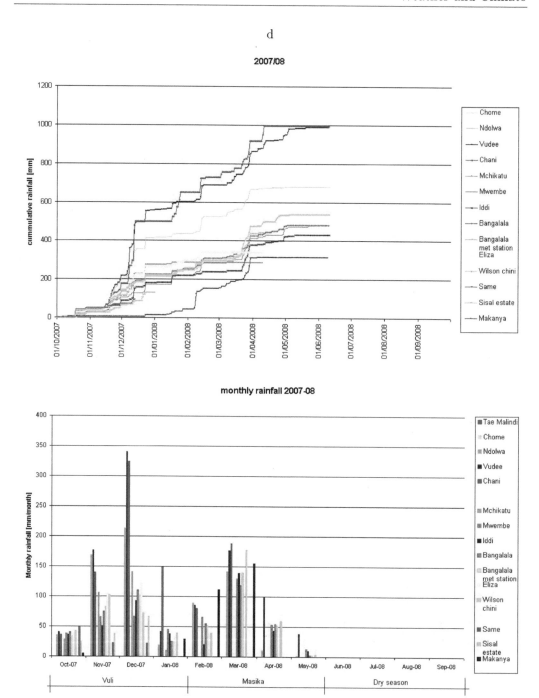

Figure 3.2 Monthly rainfall and cumulative mass curves from 13 rainfall stations in the Makanya catchment for different seasons (a-d).

As an illustration, Fig. 3.3 shows the intra-seasonal variations during the period of observation for the Bangalala rainfall station. The eight rainfall seasons, recorded during the period of study, are depicted. Only *Masika* 2006 and *Vuli* 2006/07 did not have dry spells of longer than one week, although *Masika* 2006 finished early (after 63 days, considering a growing season of 150 days, this is much too short). In this area the NGOs propose to grow a short season crop, which matures within 90 days (Makurira et al., 2007b). This would still mean that during *Masika* 2006 there is a dry spell at the end of the growing season. All other seasons experienced at least a 10-day dry spell, most of the time this was even longer. A 14-day dry spell is known to cause severe water stress that can result in a significant reduction of crop yield without supplementary irrigation (Makurira et al., 2007b; Mwenge Kahinda et al., 2007; Ngigi et al., 2005). A 21-day dry spell is considered to cause complete crop failure (Enfors and Gordon, 2007).

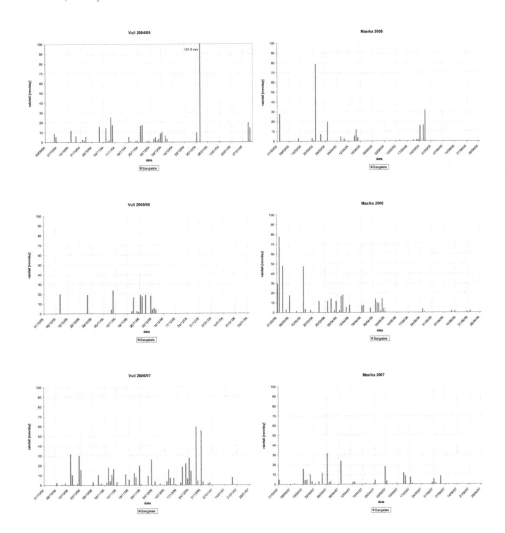

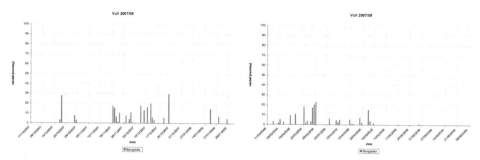

Figure 3.3 Daily rainfall during the rainy seasons for Bangalala meteorological station.

Fischer (2008) generated three dry spell probability maps for the Makanya catchment based on average seasonal rainfall within the catchment (Fig. 3.4a) and Markov chain properties, derived from the existing rain gauges network in the Pangani River Basin. The three maps indicate 80, 50 and 20 percent probability of the dry spell lengths occurring within the catchment (Fig. 3.4b-d). These maps show that the dry spell lengths in the valley of the catchment are longer than the dry spell lengths in the mountainous areas, due to the fact that the transition probabilities of the Markov chain are different in those areas. Locations with high seasonal rainfall, as occurring in the mountains, have a higher transition probability p_{01} and p_{11} (indicating the probability of a rain day (1) following a dry day (0) or rain day) than areas with low seasonal rainfall. This significantly reduced the length of a dry spell in the higher areas at a certain probability, as can be seen by Fig. 3.4 b-d.

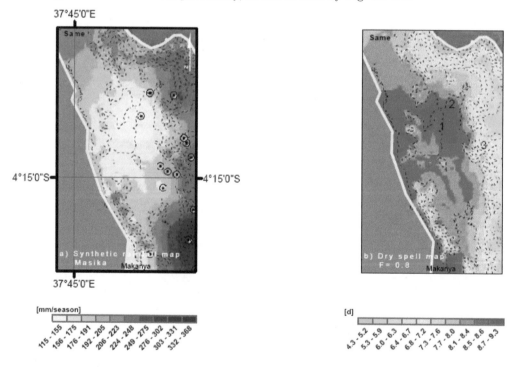

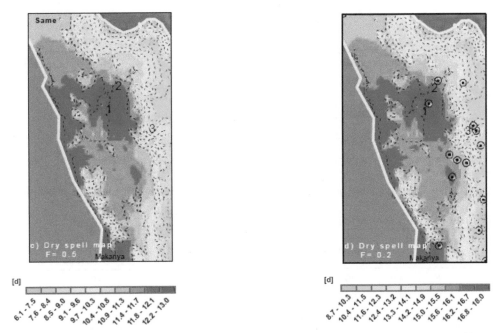

Figure 3.4 Dry spell maps for *Masika* season, a) average *Masika* rainfall, b) dry spell length 80% probability, c) dry spell length 50% probability, and d) dry spell length 20% probability (Fischer, 2008).

Besides the dry spell length, there are several other parameters affecting crop yields, such as soil properties, soil depth (D_r) and potential transpiration ($E_{t,\ pot}$). As an illustration, three locations were selected within the catchment, with estimated properties as indicated in Table 3.2. Location 1 and 2 are located in the valley of the catchment and location 3 is located in the highlands. The critical dry spell length (n_c) is computed with equation 3.1.

$$n_c = \frac{\theta}{E_{t,pot}}$$

eq. 3.1

where θ = soil moisture retention capacity of the soil (maximum available soil moisture).
The critical dry spell length for location 3 (17 days) is twice the length for location 1 (8 days), and therefore less vulnerable to dry spells. In addition, the dry spells at location 3 are shorter and occur less frequent, creating a better atmosphere for growing crops. Table 3.2 shows that in the highlands the probability of a critical dry spell is less than 10 percent for each season. On the other hand location 1 has more than 90 percent probability of receiving a critical dry spell. This shows the necessity of supplementary irrigation in the valley as Makurira et al. (2007b) indicated.

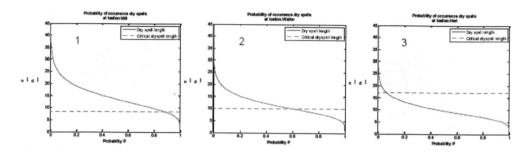

Figure 3.5 Dry spell length versus probability of occurrence for locations 1, 2 and 3 for *Masika* season.

Table 3.2 Critical dry spell length for location 1, 2 and 3.

Location	θ [%]	D_r [mm]	θ [mm]	$E_{t,\,pot}$ [mm d^{-1}]	n_{cr} [d]	Probability of n_{cr} [%] *Masika*	Probability of n_{cr} [%] *Vuli*
1	8	500	40	4.9	8.2	90	92
2	10	500	50	4.9	10.2	58	72
3	12	500	60	3.5	17.1	7	1

3.2 SPATIAL RAINFALL VARIABILITY

As large differences in topography and types of rainfall occur in the catchment, the raingauge network density is not adequate to capture the spatial variability which is apparent in the catchment (Mul et al., 2008b). Therefore, in 2006, 31 additional rain gauges were distributed to farmers in the area, who read the rain gauges every day at 9 AM (see Fig. 3.6 and Table 3.3 for locations of all rain gauges in the catchment). The dense network of rain gauges provides a good insight into the spatial variability of the rainfall in this catchment. Besides the scientific interest in this data, the farmers were also trained to analyse the data themselves which gives them more insight into the water balance of their crops.

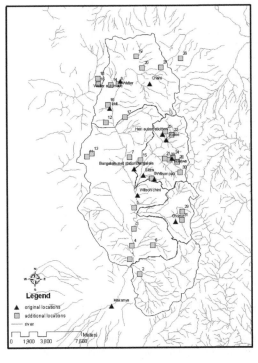

Figure 3.6 Location of all rain gauges.

Table 3.3 Additional rainfall stations in Makanya catchment.

	name	Village	altitude
1	Habibu Shabani	Makanya	734
2	Mongoloma pr. School	Makanya	763
3	Aziza Mahimbo	Makanya	773
4	Bariki Saidi	Makanya	787
5	Ramadhani Baheri	Bangalala	806
6	Salumu Bakari	Bangalala	886
7	Julius John	Bangalala	898
8	Eliapendavo	Bangalala	930
9	Juma Shabani	Mwembe	950
10	Sofia Ndani	Mwembe	971
11	Janson Hossea	Bangalala	976
12	Halima Shabani	Mwembe	977
13	Bintiali	Bangalala	986
14	Ramazani Daudi/ Marian	Mwembe	991
15	John Abdallah	Bangalala	1004
16	Nzinja	Mwembe	1014
17	Daghaseta school	Bangalala	1014
18	Juma Kasira	Mwembe	1025
19	Elianaja Nfinanga	Mwembe	1032
20	Saidi Kitunga	Mwembe	1045
21	Mwanjakiti	Ndolwa	1449
22	Nankunda Togolani	Vudee	1455
23	Amoni Mzava	Ndolwa	1537
24	Elias Mshana	Ndolwa	1538
25	Justin Mrindwa	Vudee	1539
26	Alex Baridi	Mbaga	1559
27	Nelson Imanuel	Mwembe	1560
28	Wema Kiangi	Ndolwa	1596
29	Mama Jerry	Chome	1604
30	Elineza Mwoa	Ndolwa	1638
31	Mahero pr school	Chome	1652

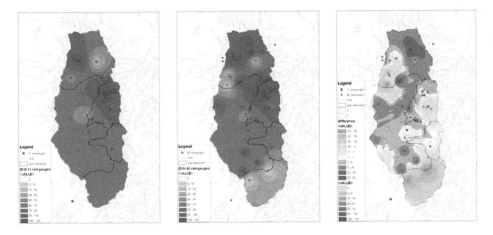

Figure 3.7 Spatial rainfall measured during 30 November 2006 event with a) 11 rain gauges, b) 42 rain gauges, and c) differences between the two interpolations in mm d^{-1}.

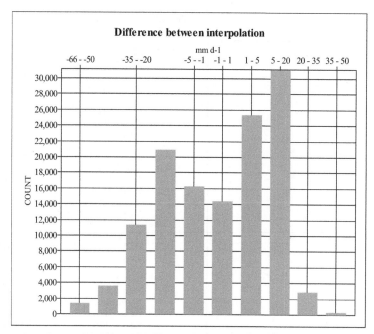

Figure 3.8 Histogram of the difference of spatial rainfall calculated with 11 and 42 rain gauges.

Based on these rain gauges a spatial map of the rainfall was made using the different sets of rain gauges. Fig. 3.7 shows the spatial rainfall distribution for the event occurring on 30 November 2006, for the 11 original rain gauges (Fig. 3.7a), and Fig 3.7b shows the spatial distribution based on all 42 rain gauges.

From a hydrological point of view the errors created by interpolation create an over- or underestimation of the spatial rainfall in the catchment, which affects the water balance. In a large part of the catchment average rainfall based on 11 rain gauges underestimates the rainfall between 0 and 20 mm d^{-1}. In the south part of the catchment average rainfall with 11 rain gauges overestimates the rainfall in this area, upto 60 mm d^{-1}. This is mainly due to the fact that the actual observations were below 15 mm d^{-1}, whereas the interpolation shows on average of 60 mm d^{-1}. Apparently, during this rainfall event the Tae area received lass rainfall than in the rest of the catchment.

3.3 POTENTIAL EVAPORATION

The meteorological station in Bangalala was installed by the SSI programme and started recording in April 2004 (Fig. 3.9). The station records on hourly basis, rainfall [mm hr^{-1}], solar radiation [W m^{-2}], air temperature [°C], relative humidity [%], wind speed [m s^{-1}] and wind direction (all mean values). At mid-night the station summarizes the data into rainfall [mm day^{-1}], maximum solar radiation [W m^{-2}], maximum and minimum air temperature [°C], maximum and minimum relative humidity and maximum wind speed. The potential evaporation is calculated on hourly basis and converted to daily and monthly values, coinciding with the manual observations at 9 AM and at midnight, using the Penman and Penman-Monteith equations (Monteith, 1965; Penman, 1948).

Figure 3.9 The meteorological station at Bangalala secondary school.

Penman's open water equation, E_{pot},

$$E_{pot} = \frac{1}{\lambda \rho} \left[\frac{s(R_n - G) + c_p \rho_a \frac{(e_a - e_d)}{r_a}}{s + \gamma} \right] \quad [\text{L T}^{-1}] \qquad \text{eq 3.2}$$

where:
λ = latent heat flux, 2,450,000 J kg^{-1},
ρ = density of water, 1,000 kg m^{-3},
ρ_a = density of moist air, J d^{-1} m^{-2},
G = heat flux density into the water body, J d^{-1}m^{-2},
c_p = specific heat of dry air at constant pressure, 1004.6 J kg^{-1}K^{-1},
γ = psychrometric constant 0.067 kPa °C^{-1}.

Vapour pressure, e_a [kPa]

$$e_a = 0.611 \exp\left(\frac{17.3 T_a}{237.3 + T_a} \right) \qquad \text{eq 3.3}$$

at temperature T_a, (in °C).

Slope of the saturated vapour pressure curve, s [kPa °C^{-1}]

$$s = 4098 \frac{e_a}{(237.3 + T_a)^2} \qquad \text{eq 3.4}$$

at temperature T_a, (in °C)

Actual vapour pressure, e_d [kPa]

$$e_d = e_a h \qquad \text{eq 3.5}$$

Aerodynamic diffusion resistance r_a [hr m^{-1}]

$$r_a = \frac{208}{u_2} \frac{1}{3600}$$ eq 3.6

Energy flux density of the net incoming radiation, R_n [J hr^{-1} m^{-2}]

$$R_n = R_{ns} - R_{nl}$$ eq 3.7

Net incoming short wave radiation, R_{ns} [J hr^{-1} m^{-2}]

$$R_{ns} = (1-r)R_s$$ eq 3.8

where:
r = albedo (0.06 for open water and 0.23 for short grass (Allen et al., 1998)) [-], R_s is observed in W m^{-2}, this is converted to J hr^{-1}m^{-2} by dividing the observed by 3600.

Net outgoing long wave radiation, R_{nl} [J hr^{-1} m^{-2}]

$$R_{nl} = \sigma(273.3 + T_a)^4 \left(\alpha \frac{n}{N} + \beta \right)\left(c + d\sqrt{e_d} \right)$$ eq 3.9

with:
α, β, c and d are climatic dependent factors.
n/N is calculated with the measured solar radiation (n) over the maximum observed solar radiation of that specific hour in that month (based on the first three years of data). It is assumed that this value is equal to the potential solar radiation at that time within that month.

The potential transpiration through the Penman- Monteith's equation then becomes:

$$E_{t,pot} = \frac{1}{\lambda\rho} \left[\frac{s(R_n - G) + c_p\rho_a \frac{(e_a - e_d)}{r_a}}{s + \gamma\left(1 + \frac{r_s}{r_a}\right)} \right] \text{[m hr}^{-1}\text{]}$$ eq 3.10

with: r_s = canopy diffusion resistance, for short grass this is 0.019 hr m^{-1}.
Fig. 3.10 shows the daily potential transpiration, calculated with the above calculation. Very high potential transpiration is calculated during the dry period in January and February, even exceeding 9 mm d^{-1}. During the colder dry period from May to August the potential evaporation is reduced to 2-5 mm d^{-1}.

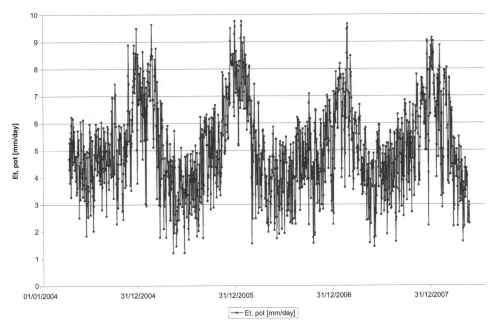

Figure 3.10 Daily potential transpiration ($E_{t,pot}$) at Bangalala meteorological station.

Fig. 3.11 shows the total monthly open water evaporation (E_{pot}) and potential transpiration ($E_{t,\,pot}$) during the period of observation (2004-2007) at Bangalala meteorological station. It shows that the evaporation is the highest in January and February, which is hot and dry. During the cold dry period evaporation is substantially lower.

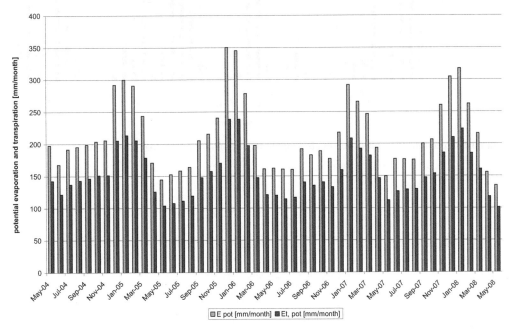

Figure 3.11 Monthly potential open water evaporation (E_{pot}) and potential transpiration ($E_{t,\,pot}$) Bangalala meteorological station.

3.4 POTENTIAL TRANSPIRATION VERSUS RAINFALL

Fig. 3.12 shows the monthly rainfall and potential transpiration measured at Bangalala meteorological station. It is very clear that the potential transpiration is much higher than the
rainfall in the area. Only three months during this period the rainfall exceeded the potential
transpiration. Although this does not mean that all rainfall is taken up as evaporation, due
to temporal variability of the rainfall within a month runoff is generated. It does mean that
most of the time water (for rainfed agriculture) is a limiting factor for crop growth. This was
also observed by Makurira et al. (2007b) for this area. It must be said that this analysis was
done using data from Bangalala, which is located in the valley of the catchment. A similar
analysis in Makanya or Same would create an even more drastic picture (less rainfall and
higher evaporation). On the contrary, comparing the rainfall and evaporation in areas such as
Vudee, would show a different picture (more rainfall and lower evaporation, because of lower
temperature). This also means that the climate in the highlands is more favourable for agriculture than in the midlands, which in turn, is better than the lowlands.

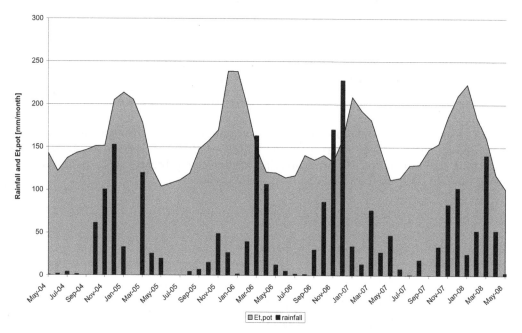

Figure 3.12 Potential transpiration versus rainfall at Bangalala meteorological station.

Chapter 4

INVESTIGATION OF FLOW SYSTEMS[6]

Understanding the hydrological flow systems in the Makanya catchment is important for predicting the impact of smallholder system innovations on the local hydrology. In Chapter 2, it has been mentioned that the geology plays a dominant role. Several techniques have been used to determine the dominant flow paths in the Makanya catchment. Geological mapping confirms the occurrence of different springs in the catchment. Hydro-chemical mapping of the springs is used to confirm the hypothesis of the flow paths, and ERT analysis to assess the locations of the saturated zone. Base flow fluctuations observed in a headwater catchment have been analysed.

4.1 HYDROCHEMICAL MAPPING

During the short rainy season *Vuli* 2005, which turned out to be exceptionally dry, groundwater and surface water sites have been sampled. A second spring water sampling was done after the *Masika* season in May 2006. The concentrations of the major anions and cations as well as dissolved silica and fluoride have been determined. A total of 30 springs have been sampled throughout the area, fifteen of the original springs were sampled twice. Electrical conductivity (EC), pH, temperature and alkalinity have been determined on site. Anions (Cl⁻, F⁻, SO_4^{2-} and NO_3^-) have been determined with Dionex Ion Chromotograph, dissolved amorphous silica (SiO_2) has been determined with HACH Colorimeter and cations (Na^+, K^+, Ca^{2+} and Mg^{2+}) have been measured on an atomic adsorption spectrometer (Perkin–Elmer model). All samples (spring samples and samples obtained from the hydrographs, see Chapter 5) have been screened by means of comparing the laboratory and field results and also the ion balance of the resultant analysis.

The ion balance, I_b [-] is defined as (Appelo and Postma, 1993):

$$I_b = \frac{\sum C_c - \sum C_a}{\sum C_c + \sum C_a}$$
eq.4.1

---

[6] This chapter is based on: Mul, M.L., Mutiibwa, R.K., Foppen, J.W.A., Uhlenbrook, S., Savenije, H.H.G., 2007a. Identification of groundwater flow systems using geological mapping and chemical spring analysis in South Pare Mountains, Tanzania. Physics and Chemistry of the Earth 32 (15-18), pages 1015-1022.

where C_c is concentration of cations and C_a concentration of anions [mol l^{-1}].

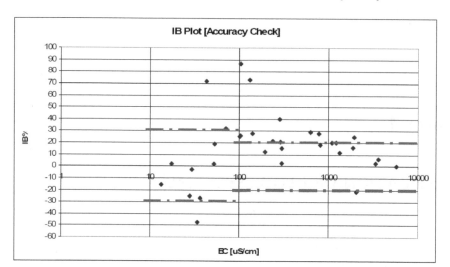

Figure 4.1 Ion balance for the dry season samples (Mutiibwa, 2006).

The ion balance could not be made for the wet season samples because the bi-carbonate was not measured in the field. A comparison of the results was done to check the consistency of the results. Moreover, the ion balance is not perfect, in general the amount of cations is over-estimated compared to the anions. Unfortunately, it is unclear where the error comes from and hence, the chemical analyses were used as they were.

4.1.1 Water Quality of Springs

In the highlands of Vudee and Ndolwa, EC values range between 18 and 500 μS cm^{-1}, while in the lowlands EC values range from 700 to 7000 μS cm^{-1}. Samples have been classified into water types, based on the chloride content, the alkalinity and the dominant cation and anion of the sample (see Stuyfzand, 1998). Zones of similar water types have been delineated on a hydrochemical map of the Makanya basin (Fig. 4.2). Although the map could have been constructed by means of classical interpolation techniques, also knowledge on geology and geomorphology has been incorporated to arrive at meaningful zones of similar water types. In general, groundwater is extremely fresh in the highlands, and becomes more brackish and contains more bicarbonate in the river valley. Dominant ions are Ca^{2+}, Mg^{2+} and HCO_3^-, while in a number of cases brackish calcium bicarbonate/sulphate type of water is encountered. In the lower part of the Makanya catchment, several springs occur along fault lines. Although these springs do not substantially contribute to the discharge in the catchment, they are interesting from a water quality perspective. Three of the sampled lowland springs have high EC-values (EC > 3000 μS cm^{-1}) and high concentrations of almost all major cations and anions, with remarkably high concentrations of F$^-$ and SO_4^{2-}, indicating that these three samples originate from a different, possibly regional, groundwater system connected to volcanic activity (Kebede et al., 2005) (see brackish areas in Fig. 4.2).

A classification based on the chloride content and the alkalinity has been used to categorise the groundwater sources. Based on this criterion and with the help of CHEMPROC® software the following descriptions have been used:

Main water type	As a function of the chloride concentration, the main water types were determined. Sources were assigned symbols from the 'G' and 'g' dominant types in the highlands to 'F' and 'f' in the lowlands (Stuyfzand, 1998).
Water types	In determining the water types, the alkalinity index was used. Sources were assigned values from '*' to '5' representing the range from "very low" to "very high" alkalinity (Stuyfzand, 1998).

Guideline definitions for the main water type and water type are summarised in Table 4.1.

Table 4.1 Guideline definitions of main water types and water types (Stuyfzand, 1998).

Main water type Range [Cl⁻ mmol l⁻¹]	Main type; Code	Water type Range [HCO₃⁻ mmol l⁻¹]	Type; Code
< 0.14	Extremely fresh; G	< 0.5	Very low; *
0.14 – 0.85	Very fresh; g	0.5 – 1	Low; 0
0.85 - 4.23	Fresh, F	1 – 2	Moderately low; 1
4.23- 8.46	Fresh-Brackish; f	2 – 4	Moderate; 2
8.46 – 28.21	Brackish; B	4 – 8	Moderately high; 3
28.21 – 282.06	Brackish – Salt	8 – 16	High; 4
> 282.06	Salt	16 – 32	Very high; 5

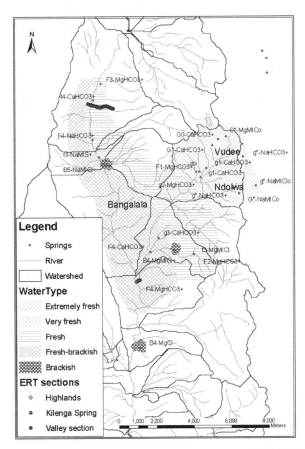

Figure 4.2 Water quality map (for typologies of the water types see Table 4.1) and location of ERT cross sections (see paragraph 4.2).

Samples have been plotted in a Ca-silicate stability diagram with anorthite as the primary Ca-silicate (see Tardy, 1971). A distinction has been made between low altitude samples and high altitude samples (Fig. 4.3). The diagram shows that silicate weathering is likely to take place, and that stable products of silicate weathering are Ca-montmorrilonite and kaolinite, whereby the former is produced in the lowland, and the latter in the highland. The differences are mainly due to variations in amorphous silica concentrations, and these high concentrations may also be the result of increased evaporation in the lowland, where average temperatures are at least 5–10 °C higher than the average temperature of the samples.

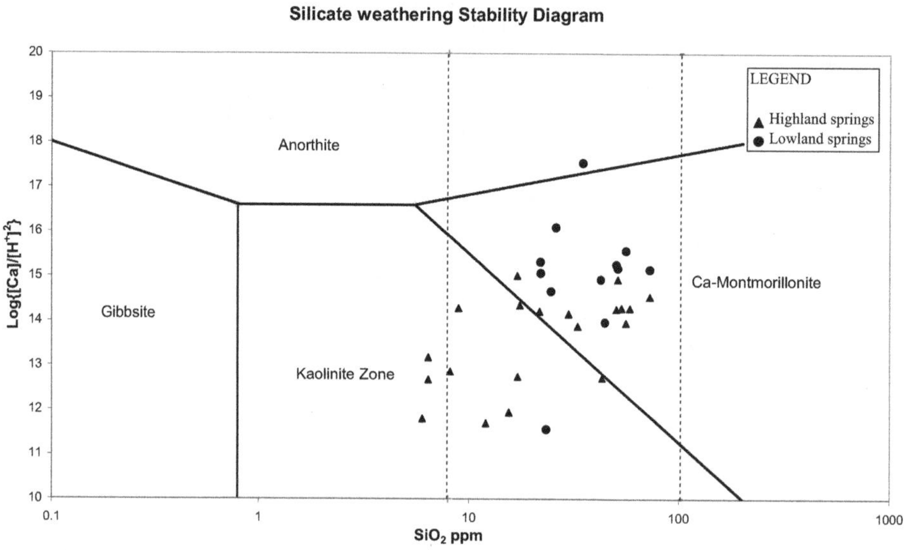

Figure 4.3 Location of groundwater samples in the silicate weathering stability diagram (after Tardy, 1971).

Inverse geochemical modelling has been conducted to determine the chemical evolution of groundwater (Kebede et al., 2005; Parkhurst and Appelo, 1999; Plummer et al., 1990). In order to do so, characteristic flow lines have been determined in the study area, and representative background and final groundwater compositions have been selected from collected water quality samples. The inverse modelling command within the hydro-chemical model PHRE-EQC[7] was used hydrochemical evolution analysis (Parkhurst and Appelo, 1999).

To obtain quantitative understanding of the occurrence of both silicate weathering and evaporation, one highland groundwater sample in Vudee has been selected as the background groundwater composition and three lowland samples near Bangalala (I, II and III; see Table 4.2) as the final groundwater composition. For the inverse modelling, it is assumed that the concentration of the major cations and anions remains constant. In general, the selection is made by knowledge of the flow system and the mineralogy along the flow path (Parkhurst and Appelo, 1999).

---

[7] PHREEQC simulates the effect of evaporation on the water samples (using the conservative parameter Chloride) and compared to the observed water quality

The results (Table 4.3) suggest that the main process explaining the difference between background and final groundwater composition can be attributed to evaporation, and the dissolution of primary silicates plagioclase and biotite due to addition of CO_2 to groundwater (Appelo and Postma, 1993; Tardy et al., 2004). The latter probably originates from decay of organic material in the soil zone. The background groundwater composition evaporates to around 10 times (from 55.4 mmol l^{-1} in pure water to 593–768 mmol l^{-1} in the final composition), while mainly plagioclase dissolves. Due to the incongruent dissolution, kaolinite "precipitates". Finally, a trace of gypsum is dissolved to account for sulphate concentrations in the final groundwater compositions of the lowlands.

Table 4.2 Concentrations for samples used in the inverse modelling.

Parameter	Samples			
	Background [Shengena]	I [Bangalala]	II [Bangalala]	III [Bangalala]
pH [-]	6.92	7.7	7.3	7.4
Ca^{2+} [mmol l^{-1}]	0.075	1.48	2.49	1.86
Mg^{2+} [mmol l^{-1}]	0.062	1.60	1.65	1.48
Na^+ [mmol l^{-1}]	0.492	2.59	1.76	1.39
K^+ [mmol l^{-1}]	0.033	0.116	0.029	0.026
HCO_3^- [mmol l^{-1}]	0.320	4.76	8.96	6.44
Cl^- [mmol l^{-1}]	0.177	1.98	1.75	0.758
SO_4^{2-} [mmol l^{-1}]	0.013	0.88	0.247	0.181
SiO_2 [mmol l^{-1}]	0.258	0.936	0.41	0.366

Table 4.3 Phase mole transfers as a result of inverse geochemical modelling using PHREEQC.

Phase	Simulations			Comments
	I	II	III	
H_2O	-5.93×10^2	-7.68×10^2	-5.510×10^2	Evaporating
Gypsum	-3.01×10^{-4}	-8.86×10^{-4}	$-6.55 \times x10^{-4}$	Dissolving
Kaolinite	$+1.58 \times 10^{-4}$	$+2.13 \times 10^{-4}$	$+1.18 \times 10^{-4}$	Precipitating
$CO_{2(g)}$	-1.78×10^{-3}	-3.70×10^{-3}	-2.50×10^{-3}	Dissolving
Biotite	-4.86×10^{-5}	-3.03×10^{-4}	-1.91×10^{-4}	Dissolving
Plagioclase	-1.94×10^{-4}	-8.84×10^{-5}	-3.27×10^{-5}	Dissolving

4.1.2 Comparison of Dry versus Wet Conditions

The spring analysis, presented above, has been done only with dry season samples (*Vuli* 2005). In this paragraph, the second set of samples (samples taken after the rainy season) is compared to the dry season samples. For both sample sets, chloride concentrations are compared to mineral concentrations (Mg^{2+}, Na^+, K^+, Ca^{2+}, SO_4^{2-} and NO_3^-) based solely on evaporation. From the graphs it can be seen that some of the parameters are below the expected line and some are above. Concentration changes derived by evaporating rain water (collected in Bangalala) have been analysed using PHREEQC software. The results of the model are plotted against the groundwater samples. Points on the line indicate that the parameter was only affected by evaporation. Points below the line indicate enrichment of the specific ions, points above the line indicate depletion of the ions. The results are presented in Fig. 4.4. The samples for the two cations of Ca^{2+} and K^+ show concentrations less than the evaporated rain samples, indicating depletion. This trend significantly increases towards the lowlands of the catchment. In contrast to Ca^{2+} and K^+, the cations of Mg^{2+} and Na^+ plot around the rain sample evaporation line (with exception of a few samples).

Comparing both parameter sets for wet and dry conditions, the trends are similar and it, therefore, can be concluded that the processes governing the spring flows are related to a slow system (Fig. 4.4). It can be observed that there is a small dilution effect in the wet season. Rainfall in a recent rainy season does not affect the concentrations of the chemicals substantially. In addition, the springs do not dry up during the dry season, base flow in both periods are very similar.

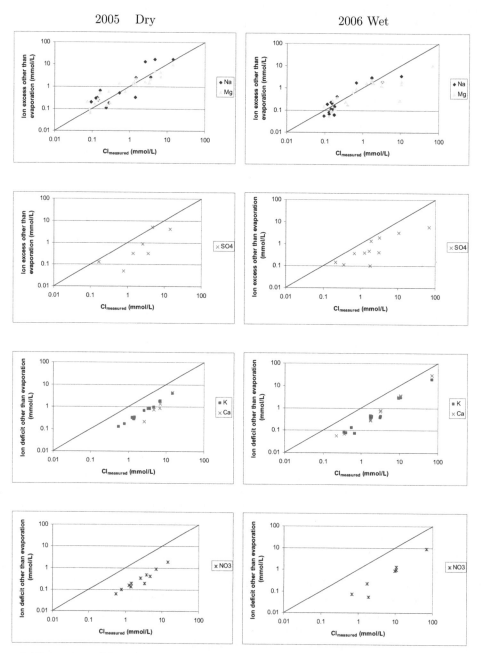

Figure 4.4 Comparison of dry and wet season samples.

4.2 GEOPHYSICAL INVESTIGATIONS

As indicated in the previous sections, the occurrence of water in the catchment is highly related to the geology and mostly dependent on a groundwater system. To map these groundwater flow systems, electrical resistivity surveys at several selected sites were conducted (see Fig. 4.2). Electrical resistivity surveys are a standard method in geology to assess the underground conditions. More recently, these techniques have been used for hydrological studies, mapping the groundwater system (Koch et al., 2008; Uhlenbrook et al., 2008; Uhlenbrook and Wenninger, 2006; Wenninger et al., 2008). Mapping the subsurface with this method relates the observed resistivity of the subsurface to geological and hydrological parameters such as rock and soil type, grain size and porosity, as well as to pore fluid properties (Loke, 2003). The determination of the subsurface resistivity is based on Ohm's law, which describes the relationship between the current density, the electrical field (voltage) and the resistivity (e.g. Loke, 2003). In order to map the electrical resistivity of the subsurface a current is induced between two electrodes and the resultant potential field is measured at two separate potential electrodes (Uhlenbrook et al., 2008). The survey configuration used in this study was an electrical resistivity tomograph (ERT), which combines surface profiling and vertical sounding into a two-dimensional (2-D) images of the subsurface resistivity (Binley and Kemna, 2005; Loke, 2003).

The resistivity surveys were carried out using the Syscal Junior Switch System with two multi-core cables with 24 electrode outlets (Uhlenbrook et al., 2008). The spacing between the electrodes varied from the sites, related to the requirements. The largest spacing between the electrodes of five meter was used, which shows the largest depth. Shorter spacing obtain a more detailed picture, with less depth. The electrodes were set along three transects, one is a cross-section across a stream in the upper parts of the catchment, the second one is a cross section in the valley, which periodically floods and used to be a wetland, nowadays it is agricultural land. The third and last transect crossed the valley for 1.5 km in an area where larger gullies dissipate in the agricultural fields. The main aim was to identify the main geological features, the area near the river bed was analysed in more detail with a shorter spacing for more detail. A dipole-dipole configuration was used as the electrical array. All measurements were carried out under similar soil moisture (dry) conditions in November 2005. The measured apparent resistivity has been processed with a 2-D inverse numerical modelling technique (software: Res2Dinv) to give the estimated true resistivities of the subsurface (for further details see e.g. Loke, 2003). In addition, surface characteristics (surface slope, vegetation, outcrops and changes in soil characteristics of the top soil) and a few hand-auger holes have been used to interpret the data. Table 4.4 gives an indication of the type of underground is associated with different resistivity values (Loke, 2003).

Table 4.4 Resistivity calibration table, based on both the drillers logs and the resistivity values indicated by the profiles.

Field calibrated values	
Resistivity [ohm m]	Formation/Rock unit
<1	Formation containing mineralised water
1 – 15	Clay: ranging from very wet, wet, moist to dry
15 – 80	Completely weathered rock: aquifer
50 – 150	Relatively weathered rock: aquifer
150 – 500	Slightly weathered and /or fractured rock: aquifer Dry loose sands
500 - 1000	Moderate to slightly fractured bedrock
>1000	Competent bedrock

4.2.1 Site I: Highlands

The first profile has been done in the highlands across one of the small tributaries. This particular site has been chosen because of the spring zones which appear on one side of the stream (left hand side in Fig. 4.5). The spring zones discharge even after the dry season period. The profile shows that the thickness of the contributing layer is about 2 meters with several pockets of groundwater, extending to depths upto 15 meters. The underlying bedrock is highly weathered near the wet locations (see resistivities between 300 to 800 Ohm m around 20 to 36 meter in the profile). At the dryer locations, less weathered rocks are apparent (resistivities of above 10,000 Ohm m). These large rocks seem to be randomly distributed along the profile and have varying shapes and the weathering processes are at various stages, linked with the wetness of the surrounding soil. The profile only shows the large rocks, but the two attempts to drill an auger well failed as smaller rocks were encountered at depths of more than 0.5 m. The shown profile indicates high heterogeneity in the profile, which has also been observed by Kessler et al. (2008) in the same area. On the right hand side, a low resistivity underground bypass is apparent which feeds the stream around the 60 meter mark in the profile. The blue pocket right under the stream could indicate a sub-surface groundwater storage, which discharges into the stream at a lower altitude.

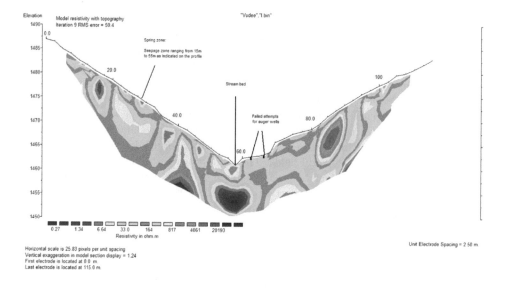

Figure 4.5 2D-geo-electrical profile across a stream in the highlands of Vudee.

4.2.2 Site II: Kilenga Spring

Along the main river in the valley of the catchment several springs occur at points where the river is cutting through alluvium (valley tilting) and the local groundwater level. One of those locations is at Kilenga spring, although these springs are perennial they do not generate a large flow and is quickly re-infiltrated in the alluvium. At this location the aquifer supplying the spring is located under agricultural fields. These fields are located in the flood plain, the aquifer is replenished by flooding from the river, which happens very rarely. The soils in this area are generally good, with good water holding capacity. Crops requiring a lot of water, such as bananas and sugar cane grow here. Deep rooting trees such as coconut and mango trees and indigenous trees indicate the existence of a deep aquifer. The profile runs across 200 meter of this agricultural land towards Kilenga spring, where the river is cutting through the aquifer and a small spring is seeping out (Fig 4.6).

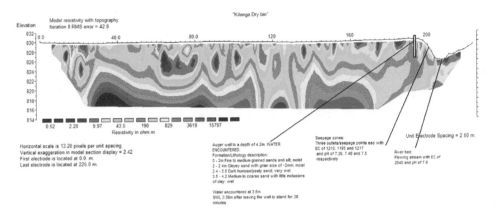

Figure 4.6 2D-geoelectrical profile at the Kilenga spring zone.

A hand-augured well 4.2 m deep has been made along the line profile to interpret the results of the geoelectrical profile. The formations encountered comprise mainly of sand ranging from fine to medium grains. Water was struck at arounsd 3.5 m and the static water level [SWL (30 minutes after construction)] was at a depth of 3.4 m. This level coincides with the spring seepage zones located close to the river bed.

Fig. 4.6 shows that the depth to the bedrock is about 10-13 meters, above which a layer is located with low resistivity values, indicating an aquifer. This aquifer is near to the surface and explains the existence of the water demanding crops and trees in the area. At a distance of 100-140 meter, from the start of the profile, a large break in the resistant bedrock is observed, which may correspond to the location of the regional fracture, which is running through the area (see Fig. 2.2). The water quality from the spring is substantially better than the water quality of the river water (EC values in the river is 2040 μS cm^{-1} and for the spring the EC is 1200 μS cm^{-1})[8]. Although the profile shows that the aquifer is predominantly local (which would assume low EC values at the spring site) this is not the case. The high EC values may be affected by mixing of the water with water from the fault line, with higher EC values.

---

[8] Sampling was done during a dry period and samples are influenced by pollution from laundry, farming activities (fertilizer etc leaching into the soil) and cattle drinking at the site.

4.2.3   Site III: Valley Cross Section

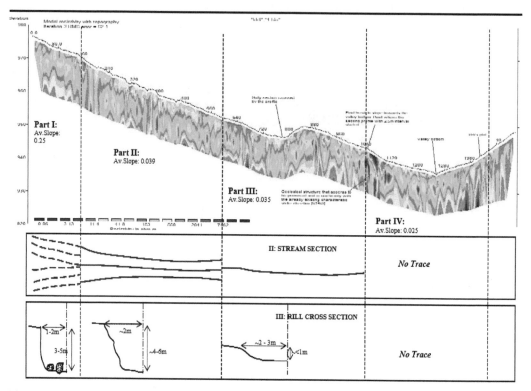

Figure 4.7 ERT measurements along the gully and dry river section (5m spacing of electrodes).

The third profile is a 1.5 km long cross section in the valley of the catchment (Fig 4.7). The profile crosses several main geological features. Throughout the profile, a hard resistant rock is apparent at 18-20 meters depth, with decreasing depths towards the river bed. On top of the bed rock, a layer with low resistivity is apparent, which comes close to the surface near the river bed. At some locations, the rock is cut through by a large fracture. From this analysis, it is not clear if water infiltrates into these fractures or if water is coming out of these fractures.

At the beginning of the profile, there are many large gullies, originating from the steep slopes upstream (slopes upto 25 percent, part I). During rainfall events water is flowing through the gullies, eroding the landscape. At some points the gullies are 4-5 meters wide and 3 meters deep. However, 500 meters downstream the gullies disappear, as water infiltrates into the alluvial aquifer. Near the stream bed water consuming crops are grown all year round and large trees, such as mango trees grow. The profile shows that at the location of the gullies the top 5 meters is a resistant layer, where little water infiltrates. This explains the large gullies, where rainwater is concentrated and deepens the existing gullies, without infiltrating into the soil, reducing the speed of the water. In part III the gullies slowly dissipate, when the top soil has a lower resistivity, water will infiltrate and the velocity of the water coming from upstream will be reduced. The gullies completely disappear arriving at the last part (no. IV), where local farmers spread the floods across their agricultural fields. In this area there are no natural springs and water only flows during and directly after a rainfall event.

The other side of the stream shows a completely different picture, the bed rock is much closer to the surface and the low resistivity layer is only located near the river bed. The aquifer below the river bed is most likely only fed by groundwater recharge from the left-hand side of the profile. In addition, surface runoff from both sides infiltrated into the river bed, by flooding the floodplain.

4.2.4 Site IV: Stream Bed

One part of the valley cross section was zoomed into. The electrode spacing is 2.5 meters focusing on the area towards the stream bed (part IV, left hand side of the river in Fig 4.7). This electrode spacing gives a maximum depth of investigation of about 7m. Although the river is currently dry in between rainfall events, this was not the case before one micro dam was built in Mwembe, upstream of the river. As mentioned before there are no springs in this area, however people dig wells near the river bed during the dry season to get water, which they strike a couple of meters below surface.

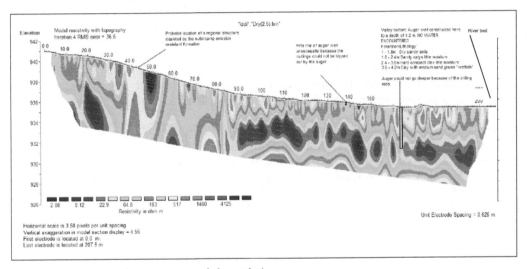

Figure 4.8 Stream bed (2.5 m spacing of electrodes).

A clear geological structure is visible. This structure may be a fault since it is accompanied by a sudden break in slope (topographically this area is in a 'U' shaped valley, 1100m into valley cross section and 40 m in stream bed profile). Similar to the valley cross section the profile shows a layer with very low apparent resistivity between 2-5 meters depth, which is an indication of layers of clay with varying moisture content (the auger well showed a thick hard pan of clay at 3 meter depth). At the deepest point of the auger hole wet-clay has been found (at the location of low resistivity in the profile).

4.2.5 Auger Wells

Several attempts were made to construct auger wells along the resistivity profiles. Out of six attempts, only two were successful and these were constructed to a depth of 4.2 m. Stable water levels (SWL) in the wells were measured 30 minutes after construction. A description of the layers penetrated for each well shown in Table 4.5.

Table 4.5 Lithological logs for the two hand-augured wells.

Well 1; streambed profile (Fig. 4.8)			Well 2; Kilenga profile (Fig. 4.6)		
Depth [m]	Observed resistivity range	Lithology	Depth [m]	Observed Resistivity range	Lithology
0 – 1.8	300 – 500	Dry sandy soils	0 – 2	43 – 190	Fine to medium grained sand and silts; Moist
1.8 – 2.4	100 – 250	Moist sandy clays	2 – 2.4	"	Clay with sand grains ~> 2 mm diameter.
2.4 – 3.6	8 – 22	Very hard compact clay	2.4 – 3.5	"	Dark humose sand; 1st water strike at 3.5 m
3.6 – 4.0	8 – 22	Clay with Medium sand grains	3.5 – 4.2	"	Medium to coarse sand grains with clay inclusions
4.0 – 4.2	8 – 22	"wettish"			SWL: 3.39m (measured 30 min. after construction)

The location of well 2 is less than 5 m from a spring zone indicated in profile II, while Well 1 is located upslope of a hand-dug well in profile IV. The "debris-flow" material was observed on the faces of the slope created by the river meandering regimes. Based on the lithological logs and the apparent resistivity values from the profiles, a calibration table including lithology and a range of expected apparent resistivity values (in the study area) along with the expected range as given by the manufacturer are indicated in Table 4.4.

To conclude, the ERT profiles confirm the geological mapping (see Chapter 2). Clear geological breaks are shown in the ERT profile in the valley are linked to the faults in Fig 2.2. In addition, the ERT profiles show that there is a substantial amount of water stored in the soil profile in the valley of the catchment. This aquifer is seasonally replenished by flooding from high intensity rainfall events. In the flood plain, local farmers use the residual soil moisture for growing bananas and sugar cane. Large trees, such as mango trees grow in these floodplains. The profile done in the highlands showed that the springs are linked to local groundwater systems, with pockets of groundwater storage. At the location of the profile, a low resistivity layer was observed under the river bed, supposedly linking more downstream to the river.

4.3 SPRING TYPES

Based on the geological mapping, combined with the hydrochemical analysis and geophysical investigations a distinction can be made between two spring types occurring in the study area. The most frequently occurring springs are those associated with the groundwater table reaching the surface. These are the only springs observed in the highlands. In most cases the springs are occurring in areas where the ridges (completely covered or outcropping) are acting as a base for the unconsolidated "debris-flow" material creating some form of damming structure (Mazor, 1991). Representative sketches of springs fed by local flow systems are indicated in Fig. 4.9.

The second type is a spring that occurs when a fault is cutting through an impermeable barriers forcing the water to discharge. This spring type is mainly observed in the lowlands and though not clearly understood, there is a considerable interaction between the shallow aquifer systems and the deep aquifer systems in terms of vertical leakage to the deeper aquifers. This could be by means of fractures in the impervious rocks or incongruent weathering.

The occurrence of several springs in the valley of the catchment with high fluoride and chloride content is associated with the main fault line observed in the catchment (Fig. 2.2). It is hypothesised that the fault line dissects through a deep aquifer with high potential, which flows out where the fault reaches the surface (Fig. 4.9).

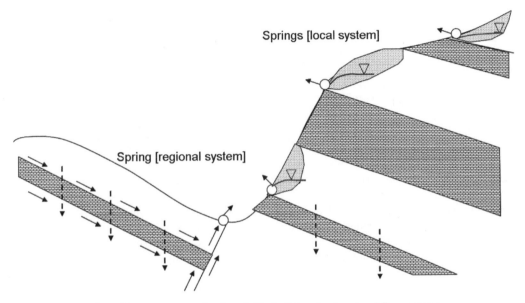

Figure 4.9 Illustration of the occurrence of springs fed by both local and regional flow systems.

The springs occurring in the highlands of the catchment are springs that discharge throughout the year, many of the local water supply and irrigation systems in the highlands rely on the water supply from these springs. The springs originating in the lowlands have a very small contribution to the water availability, next to a very local water supply function (limited as the water quality is not good) the water mainly re-infiltrates into the river bed.

4.4   SYNTHESIS: GROUNDWATER FLOW SYSTEMS IN THE MAKANYA CATCHMENT

Based on the findings from the hydrochemical and ERT analyses, a conceptual model of the catchment has been made (Fig. 4.10). The main dipping direction of the ridges is towards the eastern side of the South Pare Mountains (towards Mbaga catchment). Some water flows through fissures and faults, along the ridges, to the Mbaga catchment. The high base flow on the Mbaga side is in support of this concept (5 l s^{-1} on the Vudee side compared to 11–50 l s^{-1} on the Mbaga side). On the Makanya side, mainly isolated local flow systems occur. In this area, springs are predominantly found on the interface between debris flows from land slides and the parent hard rock, thereby discharging groundwater due to the impervious nature of the parent rock material (Fig. 4.10). The springs therefore occur at specific altitudes, corresponding with the presence of ridges, made by weathering resistant rock. This phenomenon is more pronounced in the highlands than in the lowland.

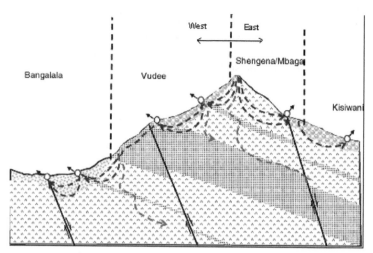

Figure 4.10 Schematisation of the predominant flow paths (Mul et al., 2007a).

Very localized flow systems prevail in the upper part of the mountains, which is reflected in the water quality of the springs. Water quality of the samples can be explained by evaporation and silicate weathering. There are, however, certain springs in the lowland, with high F^- and SO_4^{2-} levels, which cannot be explained by evaporation and silicate weathering. Therefore, it can be concluded that these samples originate from what we perceive to be a regional flow system, which is connected to regional tectonic and volcanic activities.

Although the springs in the upland appear to be coming from a local groundwater system, the flows are perennial and supply water for domestic use in the highlands as well as some water for irrigation, even during the dry season. In the lowlands, the groundwater is recharged by excess rainfall infiltrating into the floodplains and alluvial fans. In the past, the groundwater in the valley, particularly in the river bed, was also replenished by perennial flows from the highlands, which could reach Makanya village at the outlet of the catchment (Mul et al., 2006). Nowadays, the flows from the highlands have reduced in such a way that the flows do not reach Makanya village anymore (Mul et al., 2006). It is unclear what impact the reduced flows have on the recharging of the alluvium.

4.5 BASE FLOW FLUCTUATIONS

The local farmers use the base flow from the springs for supplementary and dry season irrigation. The base flow is not only affected by seasonal variation, diurnal fluctuations have also been observed in two headwater catchments. This section discusses the observed fluctuations and the hill slope processes, which could attribute to these fluctuations.

Low flow processes are generally attributed to a groundwater reservoir slowly depleting as a function of storage capacity and hydraulic conductivity. This generally leads to a recession curve which on a logarithmic scale plots as a straight line, which is observed in many cases (e.g. Fenicia et al., 2006). In few cases diurnal fluctuations were observed in the flow from small headwater catchments (Bond et al., 2002; Bren, 1997; Burt, 1979; Kobayashi et al., 1990; Tetzlaff et al., 2007). Burt (1979) showed that the soil moisture gradient during the day changed towards the soil surface, reducing the downward movement of water towards stream discharge. Kobayashi et al. (1990) showed with hydrochemical analysis that the diurnal variation can not be caused by direct evaporation in the channel, which would lead to increased concentrations in the stream. On the contrary, ion concentrations reduced with reducing water flows, which was also shown by Burt (1979). This supports the assumption of Bren (1997), that the diurnal fluctuations in flow is a consequence of riparian transpiration removing water from the phreatic aquifer in the vicinity of the stream, thereby reducing inflow to the stream. Bond et al. (2002) indicated that only a small percentage ($<$ 1 percent) of transpiration in the basin area is responsible for the diurnal fluctuations. Tetzlaff et al. (2007) indicated that because the diurnal fluctuations occur mainly during hot, dry periods, it may be a result of increased capillary tensions in riparian groundwater arising from high rates of potential evaporation restricting seepages during the day. Moreover after a long dry period, the groundwater gets disconnected from the riparian zone and the base flow fluctuations are reduced (Bond et al., 2002).

On the other hand many papers refer to groundwater fluctuations (e.g. Butler et al., 2007; Loheide et al., 2005), which they use to estimate the transpiration from the vegetation. Many advantages have been stated for using this method, such as: continuous daily estimates, integrated response, not dependent on phreatophyte types, low cost (Loheide et al., 2005). However, there are two streams of thought regarding the interpretation of the groundwater fluctuations, White (1932) initiated the idea that the total transpiration is a function of the change in storage over a period of time (shown in the fluctuations) and additional recharge of the groundwater. This recharge is visible when, during the night, the groundwater table returns to a higher level, during the absence of potential evaporation. Other researchers suggest (Bond et al., 2002) that the groundwater returning to a higher level is due to the fact that during the day trees transpire and therefore produce a tension in the soil, which allows the groundwater to be lifted as a result of capillary pressures. During the night the tension is reduced as a result of reduced water requirements of the trees, which results in reduction of the capillary rise and groundwater level increase. It is important to note that piezometers, in fact, measure the pressure variation in the phreatic aquifer, which is not necessarily the same as the variation of the water level itself.

4.5.1 Observations

Discharge has been measured in two small headwater catchments in Upper-Vudee. The two catchments have different land use, one is mainly dominated by indigenous forest and the second is mainly agricultural land, both catchments have an area of less than 1 km^2 (see Fig. 4.11). Discharge is measured with a compound weir consisting of a 90° V-notch for low flows, and a rectangular shape for the higher flows. During base flow conditions the automatic reader recorded fluctuations, which were consistent with the manual observations done twice a day (Fig. 4.12), this means it can not be a measurement error (e.g. temperature dependant measuring equipment). By coincidence the times of the manual observation are coinciding with the peaks and the lows (early in the morning and late afternoon).

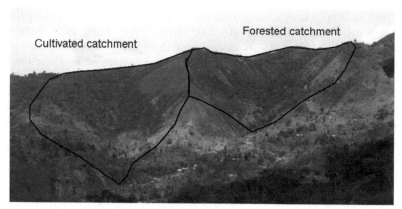

Figure 4.11 Location of paired catchments in Upper-Vudee.

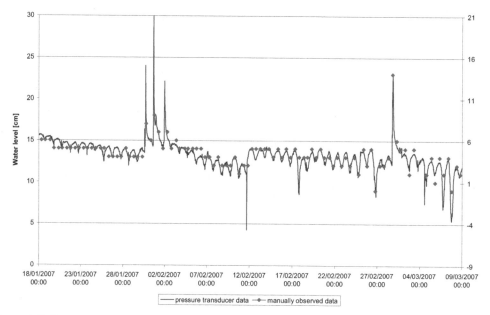

Figure 4.12 Pressure transducer data and manual observations of the diurnal base flow fluctuations in the forested catchment [cm].

These diurnal fluctuations are observed in both headwater catchments (although in different magnitudes. The forested catchment shows the largest variability, with the highest variations in the dry hot months (in this case January and February 2007). The maximum amplitude goes up to 0.5 l s^{-1} (Fig. 4.13), whereas the fluctuation in the cultivated catchment does not go beyond 0.2 l s^{-1} (Fig. 4.13). During the wet seasons (*Vuli* and *Masika*) most of the time the fluctuations are either overshadowed by rainfall events generating surface runoff or cloud cover, reducing potential evaporation. However, during rain free periods fluctuations can occasionally reappear, as can be seen in April 2007 (not shown). During the cold dry season, diurnal fluctuations are reduced to a minimum (<0.1 l s^{-1}) (not shown).

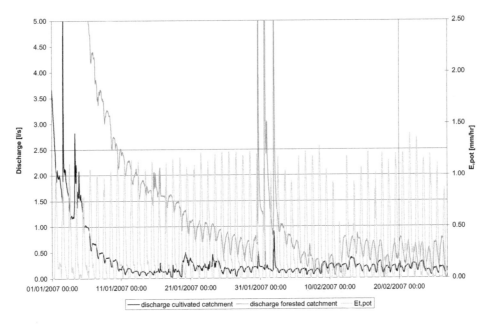

Figure 4.13 Runoff from the forested and cultivated catchment and potential transpiration, January and February 2007.

A noteworthy phenomenon is that the response of the river discharge to the increase in potential evaporation is delayed. During the dry hot period, the time lag between the peak of the transpiration and the minimum discharge is about 2 hrs, during the dry cold period the time lag is large, in the order of to 3-4 hrs. During the wet season the time lag is 1-2 hrs (see Fig. 4.14). These differences in phase shifts could be attributed to the fact that during the dry cold season the distance to the groundwater is increased, which means the suction from the vegetation has a delayed impact on the groundwater (Fig 4.15). Another possibility is that the potential evaporation is less and therefore it needs more time to build up the necessary pressure to cause the fluctuations. A third option is that only trees at a distance from the stream are still connected to the groundwater, which then cause a delayed response towards the base flow (longer flow path). The reduced correlation can be attributed to spatial differences of the vegetation and the connection to the groundwater, which is less during the cold dry period.

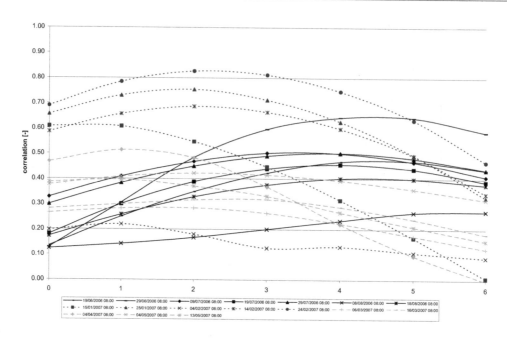

Figure 4.14 Time lag between potential evaporation and stream flow reduction.

The reduction of discharge during January and February 2007 ranges between 1 and 10 m³ d⁻¹, assuming a potential evaporation of 10 mm d⁻¹ and a K_c factor of 1.1 for forests, it would mean that between 100 - 1000 m² of forest is responsible for the diurnal fluctuations (< 1 percent of the catchment area), if all water is indeed withdrawn without flow-back.

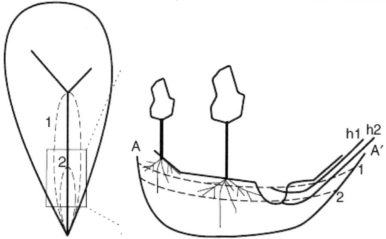

Figure 4.15 Riparian vegetation impacting base flow (Bond et al., 2002).

4.6 CONCLUSION

A dense network of springs has been observed in the highlands, which is dependent on the geology in the catchment, this is where the major part of the discharge, in particular base flow, is generated. The water quality of these springs is pristine and does not vary much, indicating a substantial and reliable groundwater supply. The riparian vegetation is using this base flow during dry periods, inducing diurnal fluctuations. During the hot and dry season in January and February, the base flow can be reduced upto 50 percent. Only riparian vegetation near to the stream induces the diurnal pattern. Increased time lag during the cold, dry season can be attributed to the lowering of the water table (see Fig 4.15).

The water resources availability in the midlands (valley of the catchment) is governed by different processes. Few springs origin in this area, which are located along fault lines and have low discharge with poor water quality. Flows from the mountain and flows generated in the catchment through high intensity rainfall events recharge the groundwater in the valley. A substantial aquifer has been observed with the ERT sections, however the quality of the groundwater may be poluted by the local springs. Groundwater samples show a high concentrations in this area. Base flow infiltrates into the alluvium and does not reach the outlet of the catchment. Large flash floods can also be generated in the midlands, as can be seen from the large gullies, which appear in the landscape. The aquifer is therefore recharged by flash floods and base flow from the highlands. Changes in the flow regime have an impact on the recharge to the groundwater, dilution the poor quality groundwater with fresh rainwater.

Chapter 5

INVESTIGATION OF HYDROLOGICAL EVENTS[9]

During the study period, three events were studied in depth to improve the hydrological process understanding in the area. Hydrochemical analysis and flood marks complemented the existing monitoring network. The first two events were relatively small floods during the *Vuli* season of 2005. The third event, recorded in March 2006, was a much larger event, which damaged part of the monitoring network. The analytical methods and the results obtained are described in this chapter.

5.1 HYDROGRAPH SEPARATION USING HYDROCHEMICAL TRACERS

Chemical hydrograph separation for two relatively small flood events has been carried out using hydrochemical tracers including electrical conductivity (EC), dissolved silica (SiO_2), and major anions (Cl^-, F^-, SO_4^{2-} and HCO_3^-) and cations (Na^+, K^+, Ca^{2+} and Mg^{2+}). The first flood event occurred on 9 November 2005 and the second on 5 December 2005, during the short rainy season. The discharge measurements and samples were taken at the weir site on the Vudee sub-catchment, downstream of the confluence of two rivers, Upper-Vudee and Ndolwa (Fig. 2). At the monitoring point, the drainage area is 25.8 km^2, with Ndolwa, draining an area of 8.4 km^2, Upper-Vudee draining 14.2 km^2 and after the confluence a catchment area of 3.2 km^2 contributes to the flow at the weir site. Separate samples were obtained from the two sub-catchments, just before the confluence during low flow. The hydro-chemical parameters have been analysed in the laboratory using an ion chromatograph and atomic absorption spectroscopy. The hydrograph separation is based on a two-component hydrograph separation, which can be described by the following set of equations:

$$c_T Q_T = c_s Q_s + c_g Q_g \qquad\qquad \text{eq. 5.1}$$

and:

[9] This chapter is based on two papers: Mul, M.L., Mutiibwa, R.K., Uhlenbrook, S., Savenije, H.H.G., 2008a. Hydrograph separation using hydrochemical tracers in the Makanya catchment, Tanzania. *Physics and Chemistry of the Earth 33, pp 151-156* and Mul, M.L., Savenije, H.H.G. and Uhlenbrook, S., 2008b. Spatial rainfall variability and runoff response during an extreme event in a semi-arid catchment in the South Pare Mountains, Tanzania. *Hydrology and Earth System Sciences Discussions 5, 2657-2685.*

$$Q_T = Q_s + Q_g \hspace{8cm} \text{eq. 5.2}$$

where c_T is the concentration at the sampling point [M L^{-3}], Q_T is the discharge at the sampling point [L^3 T^{-1}], c_s is the concentration of surface runoff (assumed to equal rainfall values) [M L^{-3}], Q_s is the runoff contribution from surface runoff [L^3 T^{-1}], c_g is the concentration of sub-surface runoff [M L^{-3}] and Q_g is the runoff contribution from the subsurface runoff [L^3 T^{-1}]. With a known concentration for sub-surface and surface runoff, the contribution from sub-surface and surface run-off can be calculated. The concentration for sub-surface runoff is assumed to be the concentration of the pre-event water at the sampling point and the concentration of the surface runoff to be similar to concentrations observed in a rainfall sample. The assumptions of this method are further discussed in Buttle (1994).

5.1.1   9 November Event

During the first event, the raingauges in Upper-Vudee and Ndolwa observed 13.5 mm d^{-1} and 7.9 mm d^{-1}, respectively. At the outlet in the valley, hardly any rainfall was recorded: 3.2 mm d^{-1}. Hourly rainfall records for this event were observed at this station, where a light drizzle started around 7 PM, continuing until 11 PM on 8 November. Rainfall in the upper part of the catchment was observed in the late afternoon. The runoff generated by this rainfall, started after midnight, reaching its peak at 2:30 AM on 9 November (Fig. 5.1). For this event, the response time is assumed to be quick enough to neglect non-conservative behaviour of the hydrochemical parameters. Fig. 5.1 shows the concentrations of selected anions and cations during the event. The discharge graph shows the rising of the flow from 5 l s^{-1} to 25 l s^{-1} with two peaks. The samples that were collected did not contain suspended sediments, which is an indication for the absence of surface runoff. Two trends in the concentrations can be observed, one group of parameters showed significant decrease in their concentrations from the beginning of the hydrograph (Ca$^+$, Mg^{2+}, HCO$_3^-$, SO$_4^{2-}$ and dissolved silica (SiO$_2$)). Others do not seem to have a clear trend and remain fairly constant (Na$^+$, K$^+$, Cl$^-$ and F$^-$). The two-component sub-surface and surface runoff separation is demonstrated using the EC values as an example (Matsubayashi et al., 1993). The results (Fig. 5.2) show that almost the entire stream flow is generated by the groundwater reservoir (95 percent). The negative values for the surface runoff may be attributed to displacement of old water stored near the canal or due to uncertainties in the method.

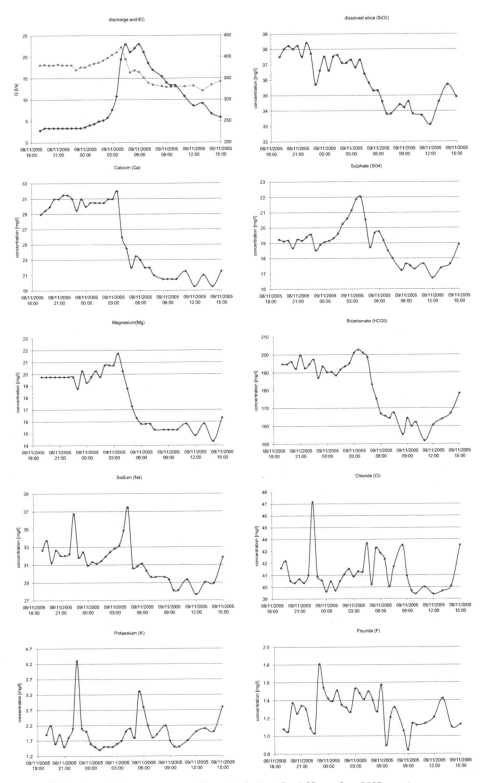

Figure 5.1 Hydrochemical parameters at the weir during the 9 November 2005 event.

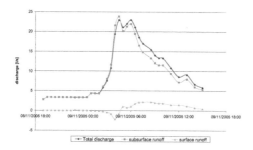

Figure 5.2 Hydrograph separation based on EC values for subsurface and surface runoff components.

Table 5.1 Observed background concentrations [mg l^{-1}].

	Upper-Vudee	Ndolwa	Weir
EC [μS cm^{-1}]	315	659	380
Ca^{2+}	28	41	29
Mg^{2+}	16	44	20
Na$^+$	29	49	33
K$^+$	1.4	2.3	1.8
HCO$_3^-$	174	316	195
SO$_4^{2-}$	30	80	39
Dissolved silica	38	38	38
Cl$^-$	39	53	42
F$^-$	1.4	1.4	1.1

The two upstream rivers have clearly different hydrochemical parameters, e.g., the background EC value of Upper-Vudee and Ndolwa are 315 and 659 μS cm^{-1}, respectively (Table 5.1). Other parameters show a similar difference except for dissolved silica. It was therefore attempted to investigate the contribution of the two sub-catchments to the total flow. This was done using the same two-component hydrograph separation, where the background concentrations are determined by the concentration of the pre-event flows in both rivers (Fig. 5.3). This method assumes that the concentration of both rivers remains the same, i.e., it neglects the surface runoff contribution. This is a reasonable assumption, as the contribution of direct surface runoff is almost none (Fig. 5.2). Fig 5.4 shows the results of this method using the EC values. It shows that the maximum discharge from Ndolwa coincides with the first peak at the weir site, while the maximum discharge from Upper-Vudee coincides with the second peak at the weir. Using EC for this analysis may be controversial for the fact that EC behaves in a non-conservative manner, therefore the same method has been applied using other hydrochemical tracers. The analysis was done using all hydrochemical parameters depicted in Fig. 5.1, whereby SO$_4^{2-}$, Mg^{2+} and Na$^+$ give similar results as the analysis based on EC. These also indicate a difference in the origin of the two flood peaks (Fig. 5.4). The analysis shows that between 81 and 91 percent of the total flow were generated by Upper-Vudee sub-catchment, which is not only the larger catchment but also received the largest amount of rainfall during this event.

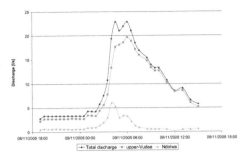

Figure 5.3 Hydrograph separation based on EC values for sub-catchments contributions.

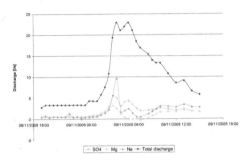

Figure 5.4 Hydrograph separation based on Mg^{2+}, Na^+ and SO_4^{2-}, showing contribution of Ndolwa sub-catchment.

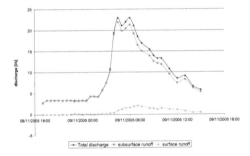

Figure 5.5 Hydrograph separation based on dissolved silica for sub-surface and surface runoff components.

The above analysis shows that it is not possible to separate the hydrograph between sub-surface and surface runoff without knowing the chemistry of the upstream situation. However, since the dissolved silica concentrations in both rivers are similar (Table 5.1), sub-surface and surface runoff contributions can be distinguished through a two-component hydrograph separation based on dissolved silica, which was also found by Wels et al. (1991) and Uhlenbrook et al. (2002). Fig. 5.5 shows the results from the application of this method, it demonstrates that almost the entire flood came from sub-surface runoff (98 percent). This confirms the assumption of negligible surface runoff contribution, which was made to quantify the contributions from the two sub-catchments.

5.1.2   5 December Event

The second reported flood event occurred on 5 December 2005, with the knowledge generated with the first event, samples were now also taken before the confluence of the two rivers, showing the chemical evolution at those two points. Fig. 5.6 shows the chemical evolution at the three points for the selected parameters. A total of 7 mm and 17.6 mm of rain was recorded by the manual rain gauge stations in Upper-Vudee and Ndolwa respectively during the period of sampling.

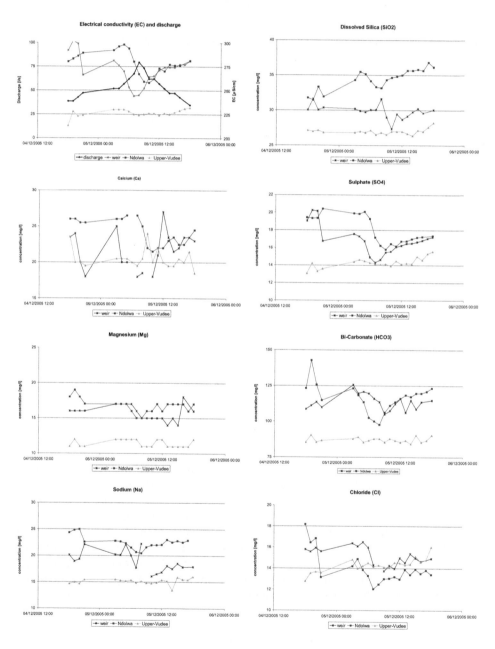

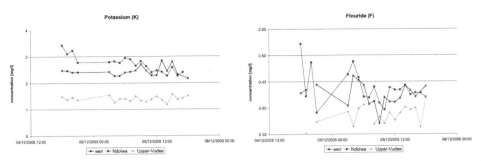

Figure 5.6 Hydro-chemical parameters at the weir and from the tributaries during the 5 December 2005 event.

The rainfall amount during this event is similar to the one observed during the 9 November flood. However the increase in discharge is much larger (40-80 l s^{-1} as compared to 5-25 l s^{-1}). This is due to the fact that this flood occurred well into the rainy season, whereby the antecedent moisture condition was much more favourable for runoff generation and the base flow was already higher than the peak flow of the first hydrograph. As a result the background hydro-chemical parameters are substantially lower. The chemical evolution in the Upper-Vudee sub-catchment is quite stable. This could be attributed to the fact that the rainfall (7 mm) in Upper-Vudee did not result in an increased runoff generation at the confluence point, at least not through surface runoff. On the other hand the chemical evolution in the Ndolwa sub-catchment shows a decrease in specific parameters, such as EC, dissolved silica, SO_4^{2-}, Mg^{2+}, HCO_3^- and Cl^-. For some parameters the chemical evolution is difficult to distinguish, because of the low concentrations. In addition a delay is observed between the confluence and weir, an experiment early 2007, showed that the travel time is in the order of 1 to 2 hours during base flow.

Hydrograph separation for surface and sub-surface runoff was done based on EC and dissolved silica, as performed in the previous section (Fig 5.7). Based on this analysis, it appears that during this event 96 (EC) to 99 (dissolved silica) percent of the runoff was contributed by sub-surface runoff. Table 5.2 depicts the calculated runoff contributions at all the sampling points, based on all the parameters analysed. The analysis shows that the average contribution from sub-surface runoff at all sampling locations is high (96, 87 and 94 percent, at weir, Ndolwa and Upper-Vudee respectively). What can be seen from this analysis, is that the surface water contribution at the Ndolwa sampling point is higher than at the other points, which is consistent with the higher rainfall observed in Ndolwa.

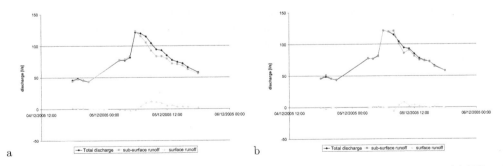

a   b

Figure 5.7 Hydrograph separation for sub-surface and surface runoff components based on (a) EC and (b) dissolved silica.

Table 5.2 Percentage sub-surface runoff based on hydrograph separation.

	Weir	Ndolwa	Upper-Vudee
EC	96	89	98
Ca^{2+}	93	85	85
Mg^{2+}	98	87	96
Na^+	94	89	94
K^+	99	90	88
HCO_3^-	95	91	97
SO_4^{2-}	91	81	98
Dissolved silica	99	98	100
Cl^-	96	76	90
F^-	99	87	97
average	96	87	94

In addition, hydrograph separation was done based on the origin (Ndolwa and Upper-Vudee), using EC (Fig. 5.8a), SO_4^{2-}, Na^+ and Mg^{2+} (Fig. 5.8b), with the base flow samples as background for the two tributaries. However for this event it may not be assumed that the surface runoff is completely negligible, particularly not for the contribution from Ndolwa. The assumption, that the base flow concentrations are representative for the concentrations in the tributaries is inaccurate in this case (see Fig. 5.7). The apparent increase of the Upper-Vudee is primarily dependent on the dilution from surface runoff anywhere in the catchment (reducing the concentrations mimicking the lower concentrations of the Upper-Vudee catchment).

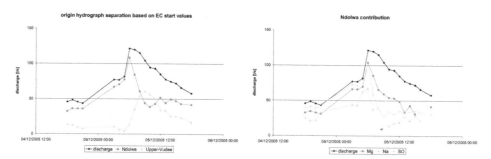

Figure 5.8 Hydrograph separation based on background concentrations of the tributaries.

Samples have been collected before the confluence of the two tributaries during the event. Deriving the contribution from each tributary to the total flow has been done using the hydrochemical analyses from these samples. The results are shown in Fig. 5.9. Determining the origin of the discharge in this case can not easily be determined through this model. One possibility could be that rainfall, generating surface runoff between the confluence and the weir influences the concentrations at the weir and, therefore, the assumption that the flow at the weir is a summation of the two tributaries is not correct (the contributing area is 12 percent of the total area).

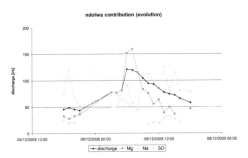

Figure 5.9 Hydrograph separation based on the chemical parameters (Mg^{2+}, Na^+ and SO_4^{2-}) of the tributaries (Ndolwa contribution).

For the 5 December event the assumption of a consistent concentration for both tributaries is not acceptable (see Fig. 5.6). Although this assumption in this case is not acceptable it does not mean it was a wrong assumption for the 9 November event, because the 5 December event was significantly bigger.

5.1.3 Discussion

Hydrochemical tracers may be readily used for hydrograph separation, however concentrations of hydrochemical tracers are a result of the processes that occur within the catchment, and, therefore, the assumption that the water quality of the fast runoff is equal to that of rainfall could be wrong and may introduce some uncertainty. This may also be the reason for the increase of some of the hydrochemical parameters at the beginning of the hydrograph. The effect on the water quality of surface runoff processes will have to be investigated. Additionally, water flowing through the unsaturated zone, diluting the sub-surface flow concentrations, may also obscure the outcomes of a two-component hydrograph separation. During both events, dominance of groundwater for flood formation was demonstrated by chemical hydrograph separations; over 95 percent was contributed by sub-surface runoff. This was confirmed by the complete lack of suspended sediments in the samples showing no erosion through surface runoff. Dissolved silica is a good tracer to distinguish between sub-surface and surface runoff, which was also found by e.g. Uhlenbrook et al. (2002); Uhlenbrook and Hoeg (2003). This outcome may not be generalised since the flood events were relatively small. Localised surface runoff may have occurred within the catchment, which re-infiltrated before reaching the stream and is therefore not visible in the chemical signature of the flood (or in sediment loads).

Changes in selected chemical parameters can be used for determining the origin of runoff (Mg^{2+}, Na^+ and SO_4^{2-}). For the 9 November event these parameters indicate that the first peak coincided with the maximum discharge from Ndolwa and the second coinciding with the maximum discharge from Upper-Vudee sub-catchment. For the 5 December event, it can be said that the major part of the hydrograph originated from Ndolwa. Further study needs to be conducted regarding hydrograph separation using isotopes.

5.2 SPATIAL RAINFALL VARIABILITY AND RUNOFF RESPONSE DURING AN EXTREME EVENT

The third event highlighted in this chapter occurred at the beginning of the *Masika* season 2006. The preceding short rainy season of 2005/06 (Oct-Jan) was extremely dry, with a seasonal rainfall amount recorded in Same of 83 mm season^{-1}, which is well below the long term average of 208 mm season^{-1}. At the weir site in Bangalala, the flow from Ndolwa and Upper-Vudee ceased completely, which according to local people happens only very rarely. This was observed earlier in 1948, 1974 and 1997 (Mul et al., 2006). Fig. 5.10a shows the weir site during this period. Upstream water allocations can reduce the flow to zero, since upstream villages have been allowed to abstract water during the day (Mul et al., submitted). During the night, however, a small trickle was observed at the weir site, when upstream villages were not abstracting.

5.2.1 Rainfall

On the first of March, heavy rainfall was recorded in the catchment, resulting in excessive flooding at the downstream end of the catchment. Water levels during the peak of the flow overtopped the gauging structure (level >1.5 m), transporting trees and big rocks, which damaged the V-notch and affected the pressure transducer post, as can be seen in the picture taken after the flood (Fig. 5.10b).

a)   b)

Figure 5.10 The weir site in February 2006 (a), and March 2006 after the flood event (b).

During the flood event in the Makanya catchment, rainfall was monitored at 14 locations, with hourly records at 5 locations and the remaining recorded daily rainfall at 9 AM every day.

Daily Rainfall

The daily rainfall recorded in the catchment on the first of March 2006 ranged from 0-122 mm d^{-1} (Table 5.3). Fig. 5.11 shows the spatial rainfall variability in the catchment based on inverse distance interpolation. In the most northern part of the catchment (altitude ranging from 950-1300 m), little rainfall was recorded (only 10 -15 mm d^{-1}). At Same village, 15 km from the catchment, no rainfall was recorded at all. In the upper part of the catchment, Shengena mountains (altitude ranging from 1400-1750 m), the rainfall recorded ranged from 75-120 mm d^{-1}. The highest amount was recorded at Chome village as 122 mm d^{-1}. In the valley of the catchment (altitude ranging around 800-900 m) a similar high intensity was recorded, 60-120 mm d^{-1}. At the outlet of the catchment, Makanya village (altitude 650-700 m)

50-60 mm d^{-1} was recorded. At Tae Malindi (106 mm d^{-1}), within a period of one day an amount of rainfall was recorded that normally falls during the entire month of March. High intensity rainfall is not uncommon in the area, during the period of observation, in Same (1934-2007), 52 times rainfall of more than 50 mm d^{-1} was recorded with two instances of rainfall above 100 mm d^{-1}. At Tae Malindi station for the recorded period (1990-2006) 8 times above 50 mm d^{-1} was recorded with one instance of above 100 mm d^{-1} (this event). At the sisal estate (1990-2006) 6 times rainfall above 50 mm d^{-1} was recorded.

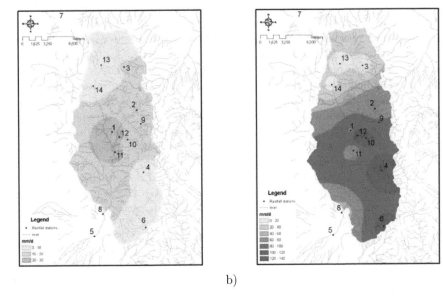

a)   b)

Figure 5.11 Spatial rainfall on 28 February (a) and 1 March 2006 (b) in the Makanya catchment.

Table 5.3 Daily rainfall in Makanya catchment, fallen on 1 March 2006, and return period based on Same data (1934-2006, 882m).

		Altitude m	Rainfall mm d^{-1}	return period (annual) yrs	return period (seasonal) yrs
1	Bangalala automatic	938	77.7	5.5	9.5
2	Vudee	1396	60.6	2.5	3.5
2	Vudee automatic	1396	82.1	7	13
3	Chani	1306	9.3	<2	<2
4	Chome	1664	122.5	67	195
5	Makanya	640	56.4	2	3
6	Tae Malindi	1417	106.0	26	64
7	Same	882	0	N/A	N/A
8	Sisal estate	698	50.3	<2	2
9	Ndolwa	1544	82.7	7	14
9	Ndolwa automatic	1544	77.5	5.5	9.5
10	Mchikatu	885	114.2	41.5	110
10	Mchikatu automatic	885	102.7	22	50
11	Wilson chini	834	63.8	2.8	4
12	Eliza	870	118.9	55	150
13	Mwembe	975	15.0	<2	<2
13	Mwembe automatic	975	15.4	<2	<2
14	Iddi	960	14.2	<2	<2

Rainfall variability in the catchment cannot be explained solely by topography. Areas with similar altitude, e.g. Mwembe and Bangalala, differed by a factor of 5, and Chani and Vudee, differed by a factor of 6-7. In general, it can be said, that the storm passed through three out of four sub-catchments, namely Vudee, Chome and Tae. In the Vudee sub-catchment the highest rainfall was recorded in the valley, above 100 mm d^{-1}, and in the upper parts of the catchment around 80 mm d^{-1}. In Chome and Tae the highest rainfall was recorded in the upper areas, above 100 mm d^{-1} and in the valley around 50 mm d^{-1}.

The highest intensity rainfall recorded during this event was of an exceptional nature. Table 5.3 shows the return period of the rainfall event based on the only available long data set in Same (73 years) using SPELL-Stat (Guzman and Chu, 2003). The Gumbel distribution has been used on the annual maximum daily rainfall events to obtain the return period of the observed rainfall. The highest intensities of the storm correspond to a return period of about 67 years (Chome, Table 5.3). Comparing the return period to the Gumbel distribution fitted on the seasonal (*Masika*) maximum daily rainfall gives even a more extreme picture, where the rainfall in Chome corresponds to a return period of about 195 years (Table 5.3). This is due to the fact that the sample size is smaller, and on top of that most high intensity rainfall events occur during the other rainy season (*Vuli*). It shows that intensities that were locally recorded have a very low probability of exceedence. It should be said that the data series that were used for the Gumbel distribution fitting were from a station outside of the catchment and at a low altitude (882 m), therefore the return periods serve only as an indication and are most probable over-estimated.

Hourly Rainfall

At five locations automatic rain gauges recorded the storm, centred around Vudee sub-catchment (Bangalala, Mchikatu, Vudee and Ndolwa), with one located in Mwembe (see Fig. 5.11). The major part of the rainfall fell in a time span of only 3 hrs (10AM -1PM). Intensities as high as 49 mm hr^{-1} were recorded in the valley of Vudee sub-catchment (Bangalala village), where also one of the highest total daily amounts was recorded (Eliza and Mchikatu) (Table 5.3, Fig. 5.12). The automatic rain gauges confirm that rainfall in the Mwembe area was considerably less than in the Vudee sub-catchment (15 compared to 80-100 mm d^{-1}). The high spatial variability, related to the high intensity rainfall observed is not uncommon to result in flash floods (Foody et al., 2004; Gaume et al., 2004).

5.2.2   Runoff

Water levels were monitored at five points in the catchment, of which four cross-sections had a known rating curve. The locations of the discharge measurements are shown in Fig. 5.9. Only at the weir site the hydrograph could be reconstructed. The pressure transducer at the weir site recorded water levels during the rising limb, but was destroyed during the peak discharge. The peak of the water level has been estimated using flood marks. The water levels of the recession limb were recorded by local observers who took water quality samples at the same time. The observations at the other sites were difficult to quantify, as the structures were either destroyed or severely damaged; however, from the few observations a response time could be estimated.

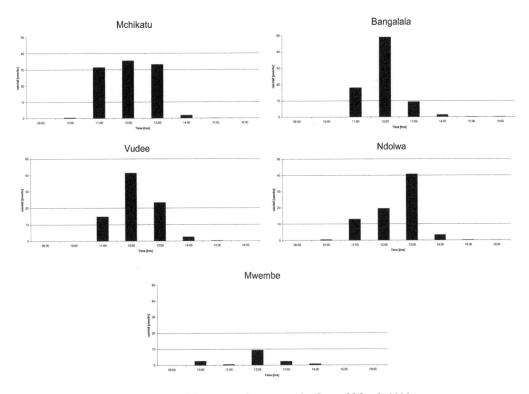

Figure 5.12 Hourly rainfall in the Makanya catchment on the first of March 2006.

Runoff at Vudee Weir

The Vudee River at the weir site drains an area of approximately 25.8 km^2. The discharge at the weir has been obtained by converting the water level into discharge using different methods. The compound weir consists of a V-notch and rectangular weir (see Fig. 5.10). Rating curves for the compound weir were obtained from Hudson (1993), and were applied for water levels below 1.5 m, above which the structure overtops (at about 13 m^3 s^{-1}). During this event the structure overtopped and maximum discharge had to be determined by converting the maximum water level to discharge. The maximum water levels upstream and downstream of the weir were obtained by surveying the flood marks. The longitudinal section is show in Fig. 5.13, with the ground level and flood marks. Cross sections were measured at 12 locations along the transect. Two techniques have been used to determine the maximum discharge, a) slope-area method and b) gradually varied flow calculations, which are described below. Similar reconstruction of the maximum discharge was done by Rico et al. (2001); Gaume et al. (2003; 2004).

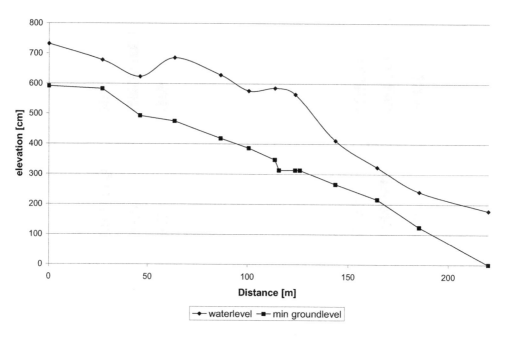

Figure 5.13 Slope of the flood marks near the weir. The diamonds indicate the highest observed flood levels (flood marks), the squares indicate the bottom level of the stream.

Slope-area Method

If we assume that near the peak the temporal deviation are small ($\partial h/\partial t = 0$) and, moreover, that the gradient of the cross-sectional area is small, then Manning's equation for permanent flow may be applied (Acrement and Schneider, 1990);

$$Q_{\max} = \frac{1}{n_m} A_f R_h^{2/3} i_e^{1/2}$$ eq. 5.3

The energy slope (i_e) is assumed to be similar to the bed slope (i_b, an average of 3 percent (see Fig. 5.13)). Only the section upstream of the weir has been used for the calculations as this is seen as a uniform stretch of the river. It is assumed that during permanent flow, the flow-through area (A_f, m^2) and hydraulic radius (R_h, m) can be obtained from the flood marks (assuming this is indicating the maximum water level). The roughness coefficient (Manning's n_m, s m$^{-1/3}$) of 0.05 is assumed (combination of cobbles and boulders (Acrement and Schneider, 1990)). The discharge estimate ranges between 30 and 60 m^3 s^{-1}, with an average of 47 m^3 s^{-1} (depending on the distance and cross sections selected, see Table 5.4). Obviously, the cross-sectional area is not uniform, hence, the method for gradually varied flow needs to be used.

Table 5.4 Calculations of $Q_{max}[m^3 \, s^{-1}]$ between 2 cross sections.

Section nr	2	3	4	5	6	7	8
2		60.0	42.3	70.5	49.2	37.0	50.4
3			N/A	54.9	38.6	29.8	41.2
4				90.3	56.9	41.4	54.8
5					48.7	42.1	58.7
6						34.6	45.6
7							33.1
8							

Gradually Varied Flow Calculation

The second method used for estimating the maximum discharge was to simulate the water depth for the entire cross section using backwater computation (Chow, 1959; French, 1986). For this purpose, the complete section upstream and downstream of the weir has been used. Two boundary conditions are needed, where the water depth is fixed: downstream of the reach and downstream of the weir (Fig. 5.14b). The boundary condition downstream of the weir is needed as an internal boundary because critical flow occurs at this point, indicated by a circle in Fig 5.14b. In the rest of the reach the flow is sub-critical during the maximum discharge. The following equations have been used to estimate the change of water depth (dy/dx) in the longitudinal section (dx of 1 m has been applied):

$$\frac{dy}{dx} = i_b * \frac{(1 - y_N/y)^L}{(1 - y_c/y)^M}$$

eq. 5.4

with y is the actual depth and L and M are two fixed parameters (3.33 and 3 respectively (Chow, 1959)). The normal depth y_N is defined by equations 5.5 - 5.7. of the cross section. The water depth approaches the normal depth in a uniform section of the river. Manning's equation is used for the normal depth:

$$A_f R_h^{2/3} = n_m \frac{Q_{est}}{i_b^{0.5}}$$

eq. 5.5

where A_f and R_h are functions of y_N and Q_{esr} is the estimated discharge. Here we simplified the measured cross sections as a trapezium, where the following equations apply for A_f and R_h:

$$A_f = (b + z)y_N$$

eq. 5.6

$$R_h = \frac{A_f}{b + 2\sqrt{y_N^2 + z^2}}$$

eq. 5.7

where b is the bottom width and z is the slope of the banks.
The critical depth y_c is the depth with the critical velocity (see Equation 5.8).

$$y_c = \left(\frac{\left(\frac{Q}{B} \right)^2}{g} \right)^{1/3}$$

eq. 5.8

where B is the width of the cross section.

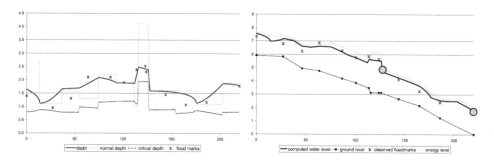

○ Fixed water level

Figure 5.14 a) Observed flood marks and modelled water level, normal and critical depth and b) Observed flood marks and modelled water and energy level, compared to a fixed reference level.

Fig. 5.14a shows the modelled water depth compared to the observed flood marks using $Q=32$ m^3 s^{-1} for the estimated discharge. The flood mark points do not fully agree with the backwater curve, which is explained by the fact that the flood marks not necessarily mark the highest water level. Branches and sticks may be pushed up, higher than the actual water level when the water hits obstacles (possibly reaching the energy level). In addition, sudden changes in the cross sections are not incorporated. Fig 5.14a shows a relatively good fit between observed and computed water depth, considering the accuracy of the flood mark observations. Fig. 5.14b shows the computed energy level (thin line) and water level (thick line) compared to a fixed reference level.

Fig.5.14a also shows the normal and critical water depth. Throughout the profile the normal water depth is above the critical water depth, and the water depth approaches asymptotically the normal depth, calculating from downstream to upstream. Only just before the weir site the normal depth is lower than the critical depth, which explains the hydraulic jump.

Hydrograph

In the hydrograph, the discharge exceeding the capacity of the compound weir but below the peak discharge has been interpolated (above the dashed line in Fig. 5.15). Between one to two hours after the rainfall started (rainfall recorded at 11 AM, fell between 10 AM and 11 AM), the discharge started to increase (7 m^3 s^{-1} was recorded at 12 PM). The maximum discharge was recorded at 12:45 PM, less than 3 hours after the start of the rains. Recession is as quick as the rise, reducing a peak flow of 32 to 4 m^3 s^{-1} within one hour.

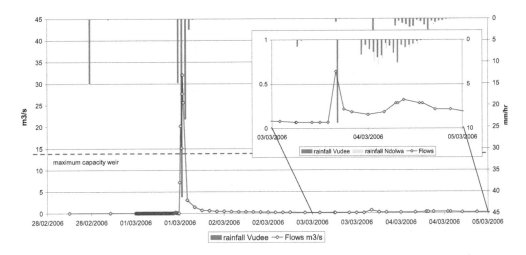

Figure 5.15 Hydrograph at Vudee weir on 1 March 2006.

The total volume of the rainfall in the upper catchment (Upper-Vudee (area 14.2 km^2), 82.1 mm d^{-1}, Ndolwa (area 8.4 km^2) 77.5 mm d^{-1} and Mchikatu (area 3.2 km^2) 102.7 mm d^{-1}), amounted to 2.15*10^6 m^3. The volume of runoff observed at the weir site, during the first 24 hrs was 0.19*10^6 m^3, which corresponds to 9 percent of the rainfall (volume under the hydrograph of Fig. 5.15; note that the uncertain estimation of the peak discharge does not significantly change the volume). As the amount of evaporation during the day itself can be assumed less than 5 percent of the recorded rainfall ($<$ 4 mm d^{-1}), the storage in the (upper) catchment should have increased substantially. This is demonstrated by the increase of the base flow observed during the subsequent period, which increased from about 15 l s^{-1} before the flood to 75 l s^{-1} in the subsequent season. The increased storage discharged an approximate 0.54*10^6 m^3 (34 percent of the rainfall) in the following season, calculation based on the outflow of a linear reservoir. The balance (57 percent) is made by evaporation in the period following the flood and percolation to the regional groundwater system (Mul et al., 2007a). Having a large part of the rainfall contributing to slow processes is not uncommon for flash floods (Belmonte and Beltran, 2001; Gaume et al., 2004), however 90 percent is very high. If this value is less, the devastating effects can be much worse.

Makanya Catchment

The two flume sites draining the two small catchments, Ndunduve and Mchikatu in the Bangalala area (2 and 5 km^2, respectively), had the quickest time of concentration, with the flumes overtopping at 9:55 AM and 10:15 AM, respectively, responding immediately to the rainfall. The site at Maji ya Chome, draining the Chome sub-catchment with approximately the same drainage area as the Vudee sub-catchment, also recorded large flows starting at 12 AM, overflowing and damaging the gauging structure. The gauge at Mgwasi at the outlet of the Makanya catchment (drainage area of 260 km^2), located at a road bridge, recorded overtopping at 12:15 AM, not much later than the occurrence of the peak flows at the sub-catchments. Concentration times at the smallest scale were almost instantaneous, at the sub-catchment scale a delay of 1-2 hrs was observed, whereas the start of the rise of flood at the sub-catchment scale and the catchment scale did not differ substantially.

5.2.3   Water Quality

Water quality samples have been taken at the main monitoring site (Vudee weir) and upstream of the confluence. The samples collected at the three sites at first had a very high
content of fine material (suspended load; diameters between 45 -100 μm). This indicates a
high contribution of Hortonian overland flow during the beginning of the flood. The small
particles can be explained by the very dry antecedent conditions, where fine particles are deposited on the top-soil and subsequently washed out by the first big rainfall event. Samples
collected after the flood peak also contained significant amounts of sediment, although less
fine particles.

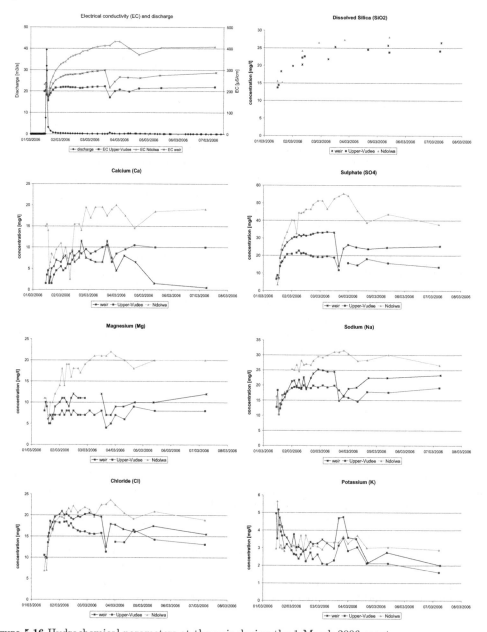

Figure 5.16 Hydrochemical parameters at the weir during the 1 March 2006 event.

Samples have been analysed for the major cations and anions (i.e. Ca^{2+}, Mg^{2+}, Na^+, K^+, SO_4^{2-}, Cl^-, F^- and dissolved silica, see for method of analysis Mul et al. (2008a)). The water quality during the flood event shows a typical pattern, except for potassium they all start with a low concentration at the peak of the event followed by a slow increase until a maximum value (Fig. 5.16). This is consistent with expectations, whereby the concentration is diluted by the surface runoff (generally with lower concentrations). Potassium shows an inverse pattern, high concentrations during the peak flows, what can be attributed to the fact that direct surface runoff picks up potassium concentrations, which indicates a significant amount of surface runoff. Similar patterns have been observed by Winston and Criss (2002); Didszun and Uhlenbrook (2008). The rainfall event, two days after the extreme event caused a decrease of concentrations in the Upper-Vudee and at the weir, but less in the Ndolwa. This is consistent with the rainfall at 4 PM, which predominantly fell in Upper-Vudee (9.4 mm hr^{-1} compared to 2.1 mm hr^{-1} in Ndolwa).

The collected hydrochemical data was used for hydrograph separation on surface and subsurface contribution. Mul et al. (2008a) showed that dissolved silica is the most appropriate parameter to separate between surface and sub-surface runoff. Unfortunately, the water quality data on dissolved silica is not suitable to do this analysis (the suspended solids affected the dissolved silica concentrations). The previous two flood events show that using EC gives similar results for estimating the groundwater contributions (Mul et al., 2008a) and several other studies have used EC as a valuable indicator (Caissie et al., 1996; Laudon and Slaymaker, 1997; Matsubayashi et al., 1993). Fig. 5.17 is the results of EC-based hydrograph separation (end member concentration for groundwater is 300 μS cm^{-1} and for surface runoff 15 μS cm^{-1}, similar to rainfall concentrations). It shows that just after the peak of the flood almost 50 percent of the runoff is generated by direct surface runoff, which is also apparent in the high concentration of suspended particles in the samples. Within 6 hours into the recession the major part of the flow is generated by groundwater (Fig. 5.17). In more studies large groundwater contributions were found during peak discharge (Frederickson and Criss, 1999; Pinder and Jones, 1969; Sklash and Farvolden, 1979). Although many others also found that with increasing intensity of the storms the event water contributions also increases (Brown et al., 1999; Caissie et al., 1996; Hooper and Shoemaker, 1986).

Hydrograph Separation

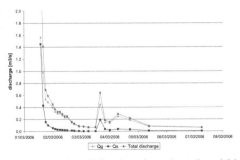

Figure 5.17 Hydrograph separation based on EC values for sub-surface (Q_g) and surface runoff (Q_s).

The hydrograph sampling at the tributaries started during the peak of the event, at the weir the samples started soon after the peak. Mul et al. (2008a) showed that the hydrochemical parameters at the weir are dependent on the runoff contributions from the two sub-catchments which are distinctly different. Therefore, hydrograph separation has been done on the origin of the flood; in other words, from which sub-catchment the flood came. In the case of the small flood events in 2005, contribution from surface water was less than 5 percent, base flow samples from the tributaries were taken as background concentrations for the hydrograph separation (Mul et al., 2008a). However, in this event surface water contributions were much higher and the background concentrations of the two sub-catchments could not be used for the hydrograph separation based on the origin. For each time step, concentrations have been obtained at three locations, these concentrations were then used to determine the contribution from each sub-catchment. Fig. 5.18 shows the hydrograph separation based on EC, Mg^{2+}, Na^+ and SO_4^{2-}.

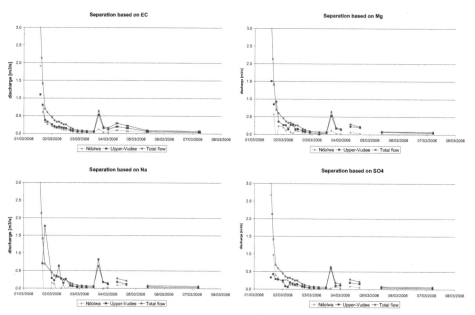

Figure 5.18 Hydrograph separation based on EC, Mg^{2+}, Na^+ and SO_4^{2-} for sub-catchment contributions.

The analysis shows that the contribution from Ndolwa (37 percent of total catchment area) during this event is in the range between 26-40 percent (Fig. 5.18), which is logical as similar rainfall amounts fell in both sub-catchments. On the other hand the small flood event on the 3rd of March was generated for a large part in Upper-Vudee, where 9.4 mm hr^{-1} was recorded as compared to 2.1 mm hr^{-1} in Ndolwa.

5.2.4 Conclusions

During this event a unique data set of an extreme rainfall and flood event was gathered in a meso-scale, semi-arid catchment. The rainfall in Vudee sub-catchment was between 75 and more than 100 mm d^{-1}, Mwembe village only recorded 15 mm d^{-1}, even though it lies less than 10 km from the centre of the storm. Furthermore, the meteorological station in Same (approximately 15 km away), with the longest record available, observed no rainfall at all during this event. Consequently, spatial variability of the rainfall during this event was quite high and not directly related to topographic features.

Reconstruction of the flow at the weir shows that the peak flow was 32 $m^3 s^{-1}$, which was reached within 1 hour of the rainfall. Only 10 percent of the rainfall passed through the weir during the event, the remainder was stored in the upstream catchment. This storage was depleted by discharge as base flow and evaporation during the following season. Hydrograph separation shows that, during the peak flows, 50 percent of the flow is contributed by direct surface runoff, followed by a recession, mainly fed by the ground water reservoir (>90 percent). During the flood event contributions from the two sub-catchments, Upper-Vudee and Ndolwa are equivalent to their respective size. During the smaller flood event on the 3rd of March contribution is mainly from Upper-Vudee, where the larger part of the rain fell (9.4 compared to 2.1 mm hr^{-1} in Ndolwa). The response time of the catchment at all scales is less than two hours. The flood caused a lot of damage to the downstream village (five houses were destroyed, the main road between Dar es Salaam and Arusha was blocked for several hours and many of the plots in the spate-irrigation system were affected), even though upstream storage reduced the flows considerably.

Extreme rainfall intensities and short concentration times characterise this event. These characteristics enhance the unpredictable nature of the floods at the outlet of the catchment. Highly localised rainfall can cause significant damage in the lower parts of the catchment, the short response time leaves little time for the residents to vacate the floodplains and bring their belongings to safety. Using the assumption of uniform spatial distribution of rainfall, the predicted runoff can easily be over- or underestimated. However, in sub-Saharan Africa the extent of the rainfall network is not adequate to capture all the spatial rainfall variability and, therefore, compromises the accurateness of the runoff predictions.

Chapter 6

HYDROLOGICAL MODELLING

Hydrological modelling was used to verify the hydrological understanding gathered from the field surveys and analyses (Chapter 4 and 5). Continuous hydrological measurements were done at several locations in the catchment with the most important ones, the weir in Bangalala and the V-notches in Vudee for the discharge measurements and the automatic raingauges in and around the Vudee sub-catchment. This chapter first describes the observed discharge and then translates the hydrological understanding into a conceptual hydrological model.

6.1   OBSERVATIONS

Runoff in the Makanya catchment is monitored at several locations (see Fig. 6.1). The main location is Vudee sub-catchment where the SSI programme together with PBWO installed a compound weir in August 2004 (Bhatt et al., 2006). At the weir location the total river flow from the two upstream catchments is measured, just before the Manoo and Mkanyeni micro dam divert the water (see further Chapter 7). The compound weir consists of a V-notch, which is able to measure low flows very accurately and a rectangular shaped weir, which can cope with large flows. This part of the river is perennial and the observer takes three times a day level measurements. In 2006, a pressure transducer started recording on hourly time steps. The observed water levels can be converted to discharge with the following equation (Hudson, 1993):

Figure 6.1 Locations of runoff gauging stations in Makanya catchment.

$$Q = 1.34 * l^{2.48} + 7.49 * \max(l - 0.305, 0)^{1.58} \qquad \text{eq. 6.1}$$

with discharge, Q, in m³ s⁻¹ and level, l in m. The second term in the equation is zero when the water level is below 30.5 cm, only the V-notch is active below this level.

The manual observations resulted in three years of daily data (Fig. 6.2). It shows that the *Vuli* season of 2005 generated extremely low runoff, to the point that the flow at the weir site completely stopped in February 2006. This had only occurred a few times before, in 1948, 1974 and 1997 (before the El Niño rains) (Mul et al., 2006). On the other hand, 2006 turned out to be an El Niño season, which reflected in the high runoff observed in November and December 2006. Both *Masika* 2005 and 2007 seasons were relatively bad seasons for rainfed agriculture (Makurira et al., 2008a), which is also reflected in the observed low flows. *Masika* 2006 started with the March 1 flood and continued with increased base flow (Mul et al., 2008b), and even with the unreliable values in April (due to the damaged V-notch), the data shows that substantial base flow was generated. This is substantially higher than normal base flow during that time of the year. The data from the *Masika* season following the March 1 flood needs to be taken with extra care, the structure was damaged (leaking in some instances) and was taken out and replaced early to mid April (this explains the fall in the level and, thus, the calculated discharge during that period). The data shows a large variability within and between seasons.

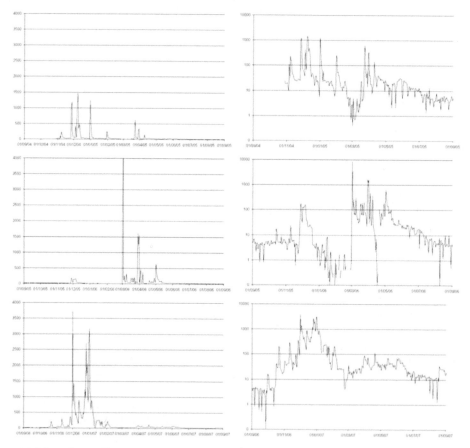

Figure 6.2 Manual runoff observed at the weir [l s⁻¹] on linear (left) and logaritmic (right) scale for 2004/05, 2005/06 and 2006/07.

Automatic observations started in May 2005, unfortunately the measurement equipment was stolen before a full time guard was employed at the weir site (February 2006). From February 2006 onwards, high resolution data (hourly rainfall and runoff data) was collected. Runoff data was affected by the 1 March 2006 flood event, only after this season could the weir be repaired (reliable data started in May 2006). From previous analysis (Chapter 3 and 5) it is clear that high resolution data is both for rainfall and runoff essential. Extremely high rainfall intensities have been recorded in the study area (> 50 mm hr^{-1}), and concentration times during single events are short. For the 1 March flood, these were between 1 and 2 hours, and for the smaller floods they were in the range of 5 - 10 hours (see Chapter 5). Daily data easily miss single hydrographs and important hydrological information. In addition rainfall intensities are crucial for runoff generation. For example, a daily rainfall event of 10 mm d^{-1} spread over 24 hours generates a completely different hydrograph than if the rainfall fell within 1 hour.

Fig. 6.3 shows the continuous runoff series, measured with the pressure transducer (hourly time step) from July 2006 to December 2007, when a second large flood destroyed the weir, temporarily ending the measurements (data collection resumed in March 2008). Comparing Figs. 6.3 and 6.2, we can see that the daily observations (although they confirm the automatic data) are not capable of capturing the hydrograph accurately. The high resolution data give much more insight into the hydrological processes; the peaks are higher; the individual hydrographs are much clearer and recession slopes can be determined with more precision. In addition, we can see that during low flows the discharge fluctuates over the day (this will be further explained in Chapter 7).

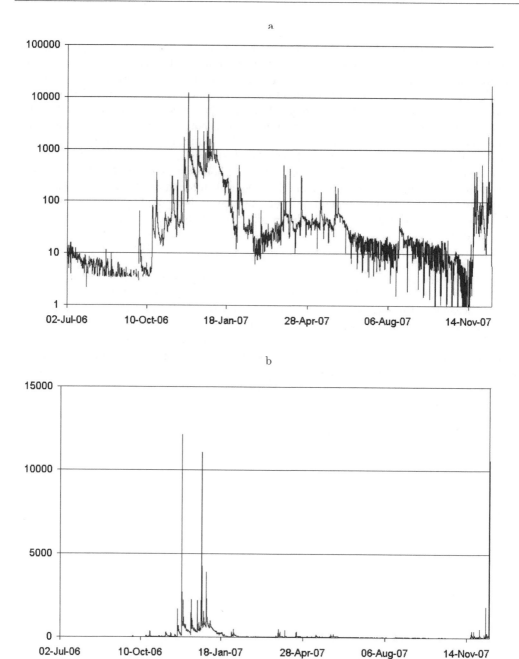

Figure 6.3 Runoff observed at weir location with pressure transducer (normal and logarithmic scale).

In the previous chapters, it has become clear that at the meso-scale, hourly data is necessary to capture all hydrological processes. Mul et al. (2008b) showed that extreme rainfall intensities (nearing 50 mm hr^{-1}) and short concentration times (less than 2 hours at the 300 km^2 scale) occur in this catchment (Chapter 5). Hence, hourly timesteps have been used to model the rainfall runoff processes in order to capture sub-daily variations. Mul et al. (2007a) showed that the geology plays an important role in the runoff generation; sloping bed rock diverts water from the Makanya catchment towards the other side of the South Pare Mountains (Chapter 4). In addition, Mul et al. (2008a) showed that for a relatively small flood, groundwater contribution is more than 95 percent. During an extreme event, surface water contribution went up to 50 percent of the peak flow, directly following the rainfall. However, the major part of the recession consisted of groundwater (34 percent of the total rainfall for the extreme event (Mul et al., 2008b)).

The above analyses are based on investigations of the larger flow system and event based investigations. Two years of rainfall and runoff data are available. Much of the hydrological processes can be derived from close investigation of the runoff data. Several important parameters can be derived from this, such as the time of concentration, runoff coefficient and the residence time (the time scale K).

Fig. 6.4 shows a selection of the data set (*Vuli* 2006), which was a good season for agriculture. What is of interest, is the fact that at higher discharge the recession slope is less steep than at lower discharge. This is remarkable as a system normally has only one type of recession, whereby the discharge is the ratio of the storage volume to the time scale of the storage (the residence time K) (e.g. Nyagwambo, 2006; Winsemius et al., 2006). In this case, at high discharge, a second storage system contributes to discharge, keeping the discharge at a higher level (K=175 days). If this were not the case, the discharge would return to a lower value shortly after a sharp peak, with K=50 days. For the discharge to remain at a high level with a slow recession, there must be an additional storage discharging into the river. It appears that once this second system is activated (near a threshold of 250 l s^{-1}) it continues for an extended period of time (this has been observed between 30 November 2006 – 19 January 2007). This phenomenon was not observed in *Masika* 2007, which was a relatively dry rainy season, while flows in *Vuli* 2007, until the weir was destroyed, also did not reach the threshold of about 250 l s^{-1}. It is, therefore, not clear if this was an exceptional season, or if this phenomenon is more frequent.

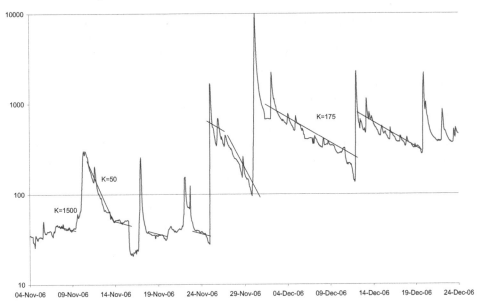

Figure 6.4 Selection of the *Vuli* 2006 season.

One possible hydrological explanation is described below. As is explained in Mul et al. (2007a), the geology of the Pare Mountains plays an important role in the hydrology. A part of the rain falling in the Makanya catchment flows towards the East of the catchment, through fractured and layered metamorphic hard rock. It is hypothesised, that the storage of the fractured rock is responsible for the slower recession and increased discharge. Above a certain threshold the water stored in the rock is connected to the debri-flow (colluvial) system and discharges towards the river (Fig. 6.5). This continues until the rock storage is again disconnected from the debri-flow system.

At the same time, the layered hard rock system drains towards the neighbouring Mbaga catchment in the East (Mul et al., 2007a), which continues to deplete the aquifer. As a result, this aquifer system will only start flowing again after the water, lost to Mbaga, is replenished and the threshold for connection to the colluvium is again exceeded.

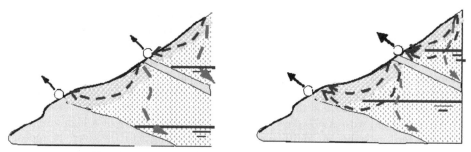

Figure 6.5 Schematisation of contributing groundwater flow systems.

6.2 METHODOLOGY

A conceptual hydrological model has been developed to test the hypothesised hydrological processes and to quantify the amount of water flowing towards the other catchment. The Lumped Elementary Watershed (LEW) approach (after Winsemius et al., 2006) has been adapted. LEW is a semi-distributed conceptual modelling approach. The river basin is subdivided into smaller watersheds, which are modelled in a lumped manner. The outlets of individual watersheds are located at confluence points or measuring locations. Three sub-catchments have been delineated: Upper-Vudee and Ndolwa sub-catchments draining the two areas upstream of the confluence, whereas the Bangalala area is the area which drains a small area at the foot of the mountains after the confluence (see Fig. 6.6). The original conceptual LEW model was developed for a large catchment (~500,000 km^2) on a monthly time step. This concept has been adjusted for the Vudee catchment (catchment area of 25.8 km^2) on an hourly time step.

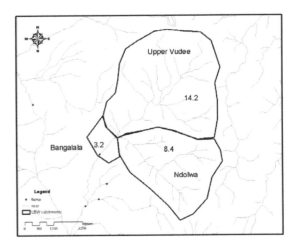

Figure 6.6 Sub-catchments defined for the LEW model.

6.3 MODEL DESCRIPTION[10]

The LEW-structure (see Fig. 6.7a) is a combination of two reservoirs representing the unsaturated (S_u [L]) and saturated zone (S_s [L]). The input into the unsaturated zone is depending on the rainfall minus the interception (fast evaporation), according to:

$$P_n = P - E_I$$ eq. 6.1

where:
P_n = net precipitation [LT^{-1}],
P = observed precipitation [LT^{-1}],
E_I = fast evaporation from interception [LT^{-1}].

On a daily time-scale, the fast evaporation is constrained by a daily threshold value D, the potential open water evaporation E_{pot} (determined by the Penman equation, see Chapter 2) and the rainfall P (Savenije, 1997; 2004), according to:

$$E_I = \min(P, D, E_{pot})$$ eq. 6.2

with:
D = daily interception threshold [LT^{-1}],
E_{pot} = open water evaporation [LT^{-1}].

At smaller time steps, however, the fast evaporation should be modelled as a small reservoir with a maximum storage capacity $S_{i,max}$ [L] that is filled by P and drained by the open water evaporation E_{pot} and that produces P_n when overflowing. Several studies show that the daily interception threshold is in the order of 5 mm d^{-1} (e.g. De Groen, 2002; Nyagwambo, 2006), which was the value adopted in this study. It is assumed, that the interception storage is depleted within one to two days.

---

[10] Derived from Winsemius et al., 2006

The unsaturated zone is modelled similar to the HYMOD model (e.g. Vrugt et al., 2002): it is assumed that the spatial variability of soil moisture capacity $S_{u,max}$ [L] can be described by a power function:

$$F_{LEW}(S_u) = 1 - \left(1 - \frac{S_u}{S_{u,max}}\right)^{B_{LEW}}$$ eq. 6.3

$$0 \leq S_u \leq S_{u,max}$$

where F_{LEW} [-] represents the fraction of the LEW that has a soil moisture capacity lower than S_u and B_{LEW} [–] determines the spatial variability of the soil moisture capacity. P_{e1} and P_{e2} [LT^{-1}] are excess rainfall components that are partitioned according to the moisture state of the catchment. When a fraction of the basin F_{LEW} is saturated, P_{e2} will only be generated over this fractional area. P_{e1} only occurs when F_{LEW}=1. The actual transpiration E_T [LT^{-1}] from the unsaturated zone S_u is described according to:

$$E_T = E_{T,pot} \min\left(\frac{S_u}{0.5 S_{u,max}}, 1\right)$$ eq. 6.4

where $E_{T,pot}$ [LT^{-1}] is the potential transpiration determined through the Penman-Monteith equation (see Chapter 2).

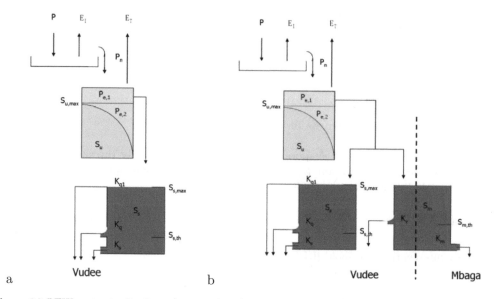

a   b

Figure 6.7 LEW conceptualisation: a) presenting the original concept (Winsemius et al., 2006), and b) the adjusted concept accounting for the slow hard rock aquifer which leaks towards the Mbaga catchment.

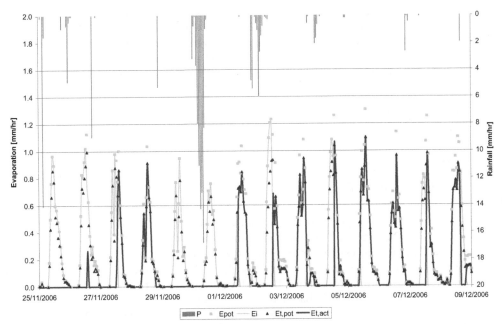

Figure 6.8 Selection of rainfall and evaporation.

Fig. 6.8 shows the calculated evaporation on a selection of the data set (the squares and triangles indicate the potential evaporation and transpiration, the lines show the calculated interception and transpiration). As can be seen from the graph, the energy used for the interception has priority over the transpiration. Once the interception storage is depleted the energy comes available for the transpiration.

The saturated zone is modelled as a linear reservoir (e.g. Fenicia et al., 2006; 2007).

$$Q = S/K$$ eq. 6.5

where S is the storage and K is the timescale of the storage.

Three runoff mechanisms for surface runoff, quick and slow groundwater drainage (K_{ql}, K_q and K_s) have been identified manually (see Fig. 6.4). Visualization of the natural logarithm of discharge provides a good indication of K_{ql}, K_q, K_s and of the threshold storage $S_{s,th}$. The first three parameters express themselves in the "steepness" of the recession curve (see Fig. 6.4), the threshold is found at the inflection point. The parameters used in the LEW are listed in Table 6.1.

6.4   OPTIMISATION

The remaining parameters in the LEW could not be determined directly from the data and needed to be defined through calibration. The Globe optimisation model (Solomatine and Dibike, 1999) has been used to optimize the parameters, B, f, $S_{m,th}$, K_m, K_v (See Table 6.1). Globe only tries to find parameters based on an optimised output. It therefore creates optimised parameter clusters, which model the time series well, but don't necessarily incorporate the characteristics of the catchment, which a trained eye can easily detect, such as the K values based on the recession curves and the threshold values at which the system changes from one state to another. By fixing these parameters and only allowing Globe to optimize the unknown parameters, this makes the model less liable to equifinality (Beven, 1993). This also makes the model more physically based and leaves the optimisation with a smaller bandwidth of options. The Nash-Sutcliffe coefficient has been used for optimisation (Nash and Sutcliffe, 1970). Equation 6.6 uses the actual values and equation 7 uses the logarithmic values.

$$R_{eff} = 1 - \frac{\sum (Q_{obs} - Q_{sim})^2}{\sum (Q_{obs} - \overline{Q}_{obs})^2} \qquad \text{eq. 6.6}$$

$$LnR_{eff} = 1 - \frac{\sum (\ln Q_{obs} - \ln Q_{sim})^2}{\sum (\ln Q_{obs} - \ln \overline{Q}_{obs})^2} \qquad \text{eq. 6.7}$$

Equation 6.6 emphasises the peak flows (as a deviation in large flows has a larger impact on the coefficient). The Nash-Sutcliffe coefficient, using logarithmic runoff values, predominantly emphasises low flows and recession curves. Globe can use both criteria for optimization, using a weighing factor. The optimization parameters which influence the peak flows have been optimized based on the normal Nash-Sutcliffe coefficient, whereas the parameters influencing the base flow have been optimized based on the log Nash-Sutcliffe coefficient. In addition, since some parameters did not show a clear optimised value, hence a step-wise calibration judging hydrograph performance visually has been applied consecutively. The obtained parameters are shown in Table 6.1.

Table 6.1 Parameters of LEW.

Parameter	Parameter	Value	Unit	Optimisation
$S_{i,max}$	Maximum interception threshold	5	mm	Literature
K_{q1}	timescale, direct runoff	1	Hour	Visual
K_q	timescale, quick flow	50	Hour	Visual
K_s	timescale, slow flow	1,500	Hour	Visual
B_{LEW}	coefficient for spatial variability of soil depth	0.9	-	Globe (general)
$S_{s,th}$	Threshold value for quick runoff	15	mm	Visual
$S_{s,max}$	Threshold value for direct runoff	5	mm	Manual
$S_{u,max}$	Maximum storage unsaturated zone	600	mm	Globe (Reff)
f	Splitting fraction	0.4	-	Topographical
$S_{m,th}$	Threshold value connecting to Vudee	8	mm	Manual
K_m	time scale, groundwater leakage to Mbaga	3,000	Hours	Globe (log Reff)
K_v	time scale, groundwater connection to Vudee	175	Hours	Visual

6.5 RESULTS

The conceptual model was first run without the additional storage in the hard rock aquifers (conceptualisation in Fig. 6.7a). Fig. 6.9 shows the results of this model. The results show clearly that the high flows in *Vuli* 2006 can not be simulated with this conceptualisation. The simulated discharge quickly falls back to lower values.

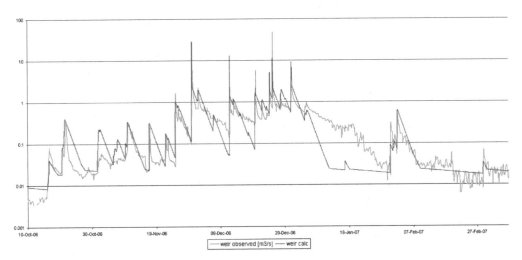

Figure 6.9 Model results for schematisation Fig. 6.7a.

The model results for the schematisation of Fig. 6.7b are shown in Fig. 6.10, it can be seen that the higher flows are much better simulated with the second schematisation. For both model runs the same parameter set is used, with only additional calibration on the additional parameters in the second storage. It should be observed that the strong variation during low flow is due to agricultural abstractions.

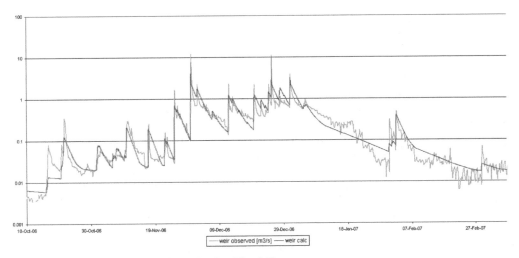

Figure 6.10 Model results for schematisation Fig. 6.7b.

That the second conceptualisation shows a better representation of the observed discharge indicates that the hypothesis (incorporated in the model schematisation) of a second storage draining towards the Vudee and Mbaga catchments is very likely to have occurred. Unfortunately, there were no water quality samples taken during this period, which could have detected a difference in water sources.

The Nash-Sutcliffe coefficient for the original schematisation is -2.49, for the improved schematisation it is 0.79. The log Nash-Sutcliffe coefficient is 0.78 for the original schematisation

(Fig. 6.7a) and 0.90 for the improved schematisation (Fig. 6.7b). This is quite good considering the high resolution of the data (hourly). It shows that the improved schematisation particularly improves the simulation of the high flows, compared to the original schematisation. The Nash-Sutcliffe coefficient is affected by the water withdrawals visible in the base flow fluctuations during the dry season (see Mul et al., 2007b), without which the goodness of fit, particularly for the log coefficient, would have been better.

As can be seen in Table 6.1, the maximum storage of the unsaturated zone is fairly large compared to the storage in the saturated zone (600 mm compared to 17 mm). The value for the unsaturated zone is related to the maximum value as used in the HYMOD distribution function, the spatially average value is therefore 53 percent less. This large storage in the unsaturated zone also acts as a buffer after a dry period, as illustrated by the first small flood in 2005, which generated a much smaller flood than a similar event well into the rainy season (Chapter 5). In addition, Kessler et al. (2008) showed that in one of the headwater catchments, recharge to the groundwater is very low, most of the storage change is located in the upper 50 cm of the soil. In contrast, the storage in the saturated zone actively involved in runoff generation is small. This storage does not include the groundwater below a runoff threshold (Winsemius et al., 2006). In addition, this value is an average over the entire catchment, while not the entire catchment area is contributing to runoff in the Makanya catchment (a large part is draining to the Mbaga side, see the fraction, f in Table 6.1). Finally, a large part of the groundwater is situated in localised debri flow systems and within cracks of the bedrock, which occupy a small percentage of the rock volume. The actual groundwater level rise is therefore much higher.

6.6 CONCLUSIONS

To be able to accurately predict runoff in a meso-scale catchment, a hydrological model with high resolution is required. At this scale, rainfall intensity is more important than daily amounts. In generating runoff, concentration times for the high intensity rainfall events are extremely short (1-2 hrs). High spatial variability of rainfall has been observed in this area affecting runoff generation (Mul et al., 2008b). To incorporate these additional complexities (and the related hydrological processes) the conceptual model has been expanded. This high resolution hydrological model has shown to have many pitfalls, the high temporal resolution data was available at only a few locations, while the spatial resolution was still not entirely sufficient (although within and near the catchment area 4 automatic rain gauges have been established).

Incorporating the observed hydrological processes improved the conceptual model substantially. So even with a limited data set, the hydrological processes in this meso-scale catchment can be understood fairly well using a multi-method approach. The additional information gained during this fieldwork proved to be very useful for constraining models and reducing equifinality of the parameters.

One of the facts which should not be ignored in this case is that the monitored catchment is far from a pristine catchment. Rainfall runoff generation is highly influenced by the land use and irrigation practices in the catchment. It is observed that during flood events sediments are washed from the surface, as a result of forest turned into agricultural land with limited attention for soil and water conservation. In addition, flash floods are observed with concentration times of less than 2 hours. As a result, infiltration and base flow have been reduced. During low flow, human activities influence the discharge with canals diverting water from the stream (Mul et al., submitted). It is therefore important to understand the human induced impact on the stream flow, in order to estimate the impact of continuing pressure on the water resources.

In addition, the model currently simulates highland processes. To model the hydrology at a larger, catchment scale, still involves major challenges. The hydrological processes in the valley of the catchment are completely different from the mountain processes. These processes need to be studied further, when upscaling this conceptual model to the catchment scale, and to predict future changes.

Chapter 7

SHARING WATER[11]

As mentioned in Chapter 2, indigenous smallholder irrigation has been practiced for a long time in the South Pare Mountains. This chapter deals with the irrigation practices in the Vudee sub-catchment in further detail. Particularly, it will focus on the water sharing arrangements within the catchment and the hydrological impacts.

7.1 WATER ALLOCATION PRACTICES

There is a renewed interest in local water allocation arrangements and, how these function. This interest is not only triggered by the steadily increasing demand for water and, hence, the growing need for better and more legitimate water allocation decisions at the local level, but also by the comprehensive water sector reforms that have occurred in many countries since the 1990s. Such reforms were often ambitious in scope, taking the national scale as a starting point, with new policies formulated, new laws enacted, new institutions established and new regulations adopted. Yet the impact on the ground has frequently been superficial, especially in Africa (Manzungu, 2004; Sokile et al., 2003; Sokile and van Koppen, 2004; Swatuk, 2005; Van der Zaag, 2005; 2007; Waalewijn et al., 2005; Wester et al., 2003). It has been argued that the disconnect between the formal statutory reality at national level and what happens on the ground may have widened rather than shrunk. Here the local level is taken as a starting point to contribute to a better understanding of why this disconnect exists and of possible ways to bridge it.

Locally developed water allocation arrangements can be surprisingly robust, as indicated by their endurance over time. This has been documented for northern Tanzania where indigenous irrigation development has a long tradition (Adams et al., 1994; Grove, 1993). A better understanding of what it is that makes such arrangements sustainable could provide new

---

[11] This chapter is based on two papers: Makurira, H., Mul, M.L., Vyagusa, N.F., Uhlenbrook, S., Savenije, H.H.G., 2007. Evaluation of community-driven smallholder irrigation in dryland South Pare Mountains, Tanzania: A case study of Manoo micro dam. *Physics and Chemistry of the Earth 32, p.p. 1090-1097*, and Mul, M.L., Kemerink, J.S., van der Zaag, P., Vyagusa, N.F., Mshana, G. and Makurira, H., 2007. Water allocation practices among smallholder farmers in South Pare Mountains, Tanzania; The issue of scale. *Agricultural Water Management, submitted.*

ideas of how institutional arrangements at larger scales could be made more effective. Such upscaling of the principles underpinning local institutional practices would contribute to bridging the identified gap through a bottom-up approach. Upscaling is likely to pose serious problems, because, beyond a certain distance between water users or user groups water allocation arrangements become increasingly sparse. Moreover, this spatial factor may have non-linear characteristics. The spatial scale, therefore, emerges as an important factor in the analysis of water institutions (see also Barham, 2001; Blomquist and Schlager, 2005; Cleaver and Franks, 2005; 2006; Swallow et al., 2001).

Water allocation practices among smallholder farmers as found within a relatively small catchment area in Tanzania are discussed at four spatial scales. (1) Between irrigators sharing one furrow system; (2) between users of adjacent furrow systems belonging to the same village; and (3) between neighbouring villages. It should be noted that agreements between distant villages (4) do not appear to exist. The possible reasons for the non-existence of water sharing arrangements at larger spatial scales will be further discussed together with the relevance of the empirical material presented in the concluding section.

7.2 AGRICULTURAL WATER USERS IN MAKANYA CATCHMENT

Makanya catchment is a semi-closed system, which only during extreme floods drains into the Pangani River. Perennial rivers originate from the Shengena Mountains on the western side of the catchment, which rise to 2,100 m. In these upland areas agriculture is practiced throughout the year, with indigenous furrows diverting water for supplementary irrigation during the two rainy seasons and full irrigation during the dry season. From a steep escarpment these perennial rivers flow into the midland areas where the water is used for supplementary irrigation during the rainy seasons. The remaining water continues its flow downhill until it reaches the valley of the catchment where the majority of the runoff is recharging the local aquifer under the sandy river bed (Mul et al., 2007a). Only flood flows, generated in one or more of the four sub-catchments reach the outlet of the catchment which the lowland farmers divert into their plots for irrigating crops such as maize and, in former times, cotton. This type of irrigation is known in the literature as spate-irrigation (e.g. Mehari et al., 2005).

There are over 100 indigenous furrows in the in Makanya catchment, each supplying water to areas ranging from 0.5 to 400 hectares. Similar to the Chagga systems described by Grove (1993), people use furrow water not only for irrigation but also for domestic use and for watering livestock. Most furrow systems are rather rudimentary in terms of materials used. Water is diverted by structures made of rocks, branches and mud. Some aqueducts exist which are made of wooden logs. The furrows are mainly hand-dug unlined small canals and sometimes stretch for several kilometres. Flood flows often destroy the intakes, which need to be rebuilt by the members of the furrow system.

Many irrigation furrows also have micro dams (75 have been identified in the Makanya catchment). These micro dams are mostly located in the upstream parts of the command area of a furrow system and serve to temporarily store water when nobody irrigates, in order to boost the diverted river flow in the furrow when farmers are irrigating. Without such reservoirs the water would not reach the most distant users because of the large transmission losses (Makurira et al., 2007a). It is important to note that many reservoirs have a fairly long history and were established by local clans before or during the colonial period, and are known by names. Over the years most reservoirs have been enlarged. More recently NGOs have assisted irrigators with lining these dams.

7.2.1 Manoo Furrow System

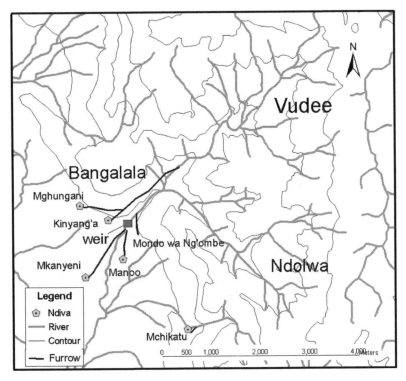

Figure 7.1 Furrow systems in Bangalala village.

The Manoo furrow system is located at the downstream end of the Vudee sub-catchment (Fig. 7.1). The furrow takes water from the river downstream of the confluence of Ndolwa and Vudee River near Bangalala. It has a total length of approximately 3.5 km (see Fig. 7.2). The main canal crosses two significant gullies. Undesired losses to lateral canals and natural drainage systems are minimised by closing off off-take points with stones and earth bunds. Not far from the beginning of the furrow is Manoo micro dam. Manoo micro dam is one of the oldest micro dams in the area which was established in 1936 by the Wadee clan. Over the years the command area and capacity increased, in line with accommodative traditional norms (Kemerink et al., submitted). In 1990, the government of Tanzania abolished natural resources management by the clanship, resulting in Manoo furrow system becoming a property of the whole community. In 2002, the micro dam was rehabilitated with the assistance of a local NGO. The new capacity of the micro dam is 1620 m^3 serving about 150 members of the community over an area of 400 hectares (Makurira et al., 2007a).

For management purposes, Manoo furrow system is divided into three zones, namely Kwanyungu (upstream), Heiziga (midstream) and Heishitu (downstream). Each zone has its own elected representative, locally referred to as Halmashauri, who is responsible for the distribution of water among its members. At the level of the Manoo furrow system, there is the water allocation committee which consists of 10 members, including the three representatives of each zone and one additional elderly advisor of each zone, a chairperson, a vice-chairperson, a secretary and a treasurer. The allocation committee meets once a week and decides which zone will receive water at which specific day. The farmers are also present at the meeting since also other issues are discussed, such as communal work and conflicts. Directly after the meeting the farmers can put in their requests for water to the representative of their zone, who will then allocate the water to some of its farmers.

After 4 PM, all water diverted from the river flows into the micro dam, which fills during the night and once filled, the water spills into the furrow downstream of the dam, which is available for whoever is interested. Obviously, the irrigators located in the most upstream zone are at an advantage to use this water, these farmers are often descendants of the Wadee clan. Distributing allocated water starts in the morning and continues until 4 PM and, during this period, diversion into the micro dam also continues. Distribution of the water to the distribution zones is managed by the representative of the zone. Farmers receiving their irrigation turn are responsible for the distribution among themselves and for opening up the bunds. The water is spread on the fields using flood irrigation. Normally up to four beneficiaries receive an allocation per irrigation turn to the zone, depending on storage available in the micro dam.

In allocating the water to the farmers the representative takes into account whether the farmer already received an allocation and whether his or her plot is well prepared to receive the water. It is very rare that a farmer gets more than two official allocations within a given season. Some farmers go without an allocation for the entire season, as the capacity of the micro dam and furrow is far too small to serve the entire command area. Usually there are no requests for irrigation when rainfall is sufficient, but a week or more after a dry spell has started, when crops start experiencing water stress, many requests are received at the same time. However, during such periods river flows may also be low, which means less water available to allocate. In case flows are extremely low, farmers in the upstream zone get priority. The justification given by irrigators is that this is because of the high transmission losses in the system, and that small water releases would not go far into the furrows. Allocating the water nearer to the source is, therefore, perceived to be more efficient. Another explanation is that the farmers in this zone are the descendents of the Wadee clan which originally started the Manoo furrow system and, therefore, may have the strongest claims to furrow water (Kemerink et al., submitted).

Water Requirements Manoo micro dam

Calculations based on management practices shows that the equivalent size of the micro dam ($1,620$ m^3), which is used to supply a command area of 400 hectares, does not make a substantial difference to the livelihoods of the members. Full-scale irrigation is out of the question as $2 * 10^6$ m^3 of water would be needed to grow maize, the preferred crop, which has a water requirement of 500 mm season^{-1}, but supplementary irrigation is also not fully feasible for the entire command area. The allocation schedule in practice aims at providing each farmer every 10 days with water, this is not achievable as, practically, only 4 out of the 150 members receive water during an allocation day. With the micro dam receiving water for three days only in a week (see paragraph 7.3), it would take 12.5 weeks before the first farmer receives a second allocation. As a result the allocation schedule, given its reservoir size and demand, is not effective for dry spell mitigation for the entire command area.

Crop water requirements were calculated using data collected during the short rains (*Vuli*) of 2004/05. Evaporation data used, was obtained from a nearby class A evaporation pan (E_{pan}). A pan factor $K_p = 0.7$ was used to determine the reference evapotranspiration (E_0). Crop coefficients (K_c) were used to compute the actual crop water requirement (Allen et al., 1998). The crop studied was maize, which is the most preferred crop in the area. The calculated crop water requirement was compared with the rainfall (P) on a monthly basis from which the required supplemental irrigation was derived for the entire area. Losses between the point of release and use points were not factored into the calculations. The results obtained are shown in Table 7.1. The table simplifies the situation as conditions pertaining to a given month also influence conditions in the preceding month, e.g. rainfall in November may evaporate in December. Rainfall distribution in a month is not uniform and will not be used by the plant uniformly and that residual soil moisture levels are ignored in this case. Also, it may not be the case that the excess water in a given month will be available for use in the subsequent month.

Table 7.1 Crop water requirement for maize (short rainy season 2004/2005).

Month	E_{pan} [mm month^{-1}]	E_o [mm month^{-1}]	K_c	E (CWR) [mm month^{-1}]	Rainfall [mm month^{-1}]	Deficit/ surplus [mm month^{-1}]	Area (ha)	Irr. Req [10^{3}m^3 month^{-1}]
October	115	80.3	0.5	40.1	41.8	-1.7	400	- 6.6
November	117	81.6	0.85	69.3	118.6	-49.3	400	- 197.1
December	177	123.8	1.2	148.5	51.4	97.1	400	388.4
January	211	148.0	0.9	133.2	172.0	-38.8	400	- 155.3
February	226	158.2	0.6	94.9	0.0	94.9	400	379.7
per season	**845**			**486.1**	**383.8**	**102.3**		**409.1**

The maximum capacity of Manoo micro dam is 1,620 m^3. According to calculations in Table 7.1, the total seasonal supplemental irrigation requirement was 1,021 * 10^3 m^3. Therefore, Manoo micro dam was required to fill 630 times to suffice seasonal requirements, assuming 100 percent efficiency. Only the month of November received sufficient rainfall, with the rest of the months requiring supplemental irrigation, including at crucial growth stages. The irrigation water requirement for December was 124 mm month^{-1}, equivalent to 495 * 10^3 m^3 month^{-1} for the 400 hectares without taking into account irrigation efficiency. This implies that Manoo micro dam needed to be filled 305 times, or an average of 10 times a day, to cater for the deficit in crop water requirements. Practically, this is not possible as it takes 2–4 h to fill the dam, depending on river flow, which would likely be low during this period as well.

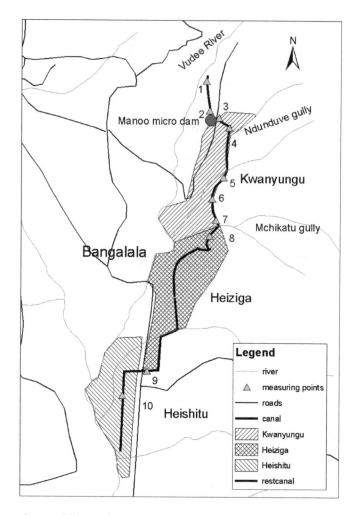

Figure 7.2 Command area of Manoo furrow system and discharge measurement locations.

Table 7.2 Location measuring points.

Station number	Type of gauge	Distance between stations [m]	Cumulative distance [m]	Remark
1	V-notch	0	0	River diversion
2	V-notch	525	525	Silt trap
Manoo micro dam				
3	H-flume	0	0	After Manoo micro dam
4	V-notch	135	135	
Ndunduve gully				
5	V-notch	514	649	
6	V-notch	296	945	Diversion to farm plot in Kwanyungu zone
7	V-notch	130	1,075	
Mchikatu gully				
8	V-notch	49	1,124	
9	V-notch	1,295	2,419	Before crossing main road
10	H-flume	660	3,079	Farm plot in Heishitu zone

Manoo irrigation system efficiency

From this analysis it shows that the micro dam as a storage structure is not very effective to supply sufficient water for all members of the water user group. The membership of the furrow system is clearly too large for the available water. Moreover, new members are located at unfavourable locations in the furrow system, i.e. at the tail ends of the canals. In addition, the estimated water requirements are not taking into account the distribution losses in the distribution canals. At four instances discharge was measured in the distribution canals, in order to estimate the distribution losses. Fig. 7.2 shows the location of the measuring points. The conveyance losses from the outlet of Manoo micro dam to the farmers' plots were estimated to range between 75 and 85 percent. This was determined by comparing the release from the micro dam with the amount of water that reached the field. Measurements at locations along the main canal (see Fig. 7.2; Table 7.2) indicated losses ranging between 10 and 20 percent between two measurement points, with 14 percent loss recorded along the diversion canal from the river into the micro dam. Losses were higher in lateral canals due to poor maintenance, steeper slopes and shallow banks which led to spillages in some cases. Since allocations to each farmer occur only once per season, it may be expected that the canal bed would be very dry prior to a release, hence conveyance losses are expected to be much higher in the lateral canals.

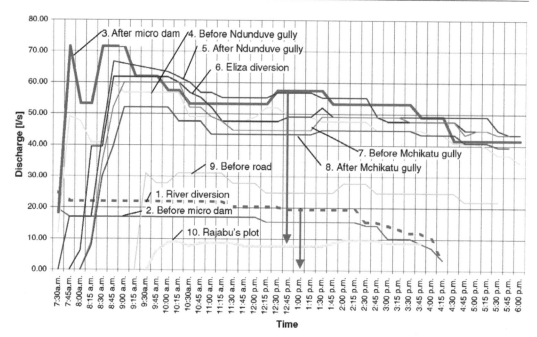

Figure 7.3 Discharges measured at different measuring points in the main canal, 24 April 2005.

Fig. 7.3 shows the results of discharge measurements for one of the allocation days at different measuring points from the dam down to the irrigated sites located 3 km downstream of the dam. The releases from the micro dams started at 07:30. Measurements were taken until 6 PM. Diversion from the river ended at 4 PM. The average discharge at the diversion point from the river into the micro dam was 18.2 l s^{-1} while the average release from the dam was 51.6 l s^{-1}. The average discharge reaching the allocated farm (Rajabu's plot in Heishitu zone) was 8.3 l s^{-1}. During the observation period 84 percent, (43.3 l s^{-1}) of the water released from Manoo micro dam did not reach the allocated farm, this figure is based on the steady state situation reached after 2.5 hours. With the diversion from the river being only 18.2 l s^{-1} it is highly improbable that the water would have reached the farmer's plot if the micro dam had not been there. This indicates that, while the size of the micro dam is considered to be too small, the structure acts as a night storage reservoir which helps to increase the flow in the furrow system thus allowing water to reach lower parts of the scheme. An interesting observation is, that where the canal follows the Ndunduve gully for 300 m, initially it takes the water 30 min to reach the point where the canal continues, after which the flow has increased by about 14 l s^{-1} (see Fig. 7.3). This is explained by the presence of a natural spring which discharges into this gully and contributes to the flow in the canal. The second gully had no impact on the discharge, as the canal crosses, but does not follow this gully.

7.2.2 Makanya Spate-irrigation System

At the outlet of the catchment near Makanya village, a spate-irrigation system was established around 1967. It consists of 6 diversion canals, namely Suji Kitivo, Kwa Sued, Mwembe mmoja, Kimbau/Maganda, Kuku and Wanda (Mutabazi et al., 2005). The intake structures divert the flood flows coming from the South Pare Mountains to irrigate the land. The soils are alluvial deposits which are nutrient rich and have a large water holding capacity. This means that in most years two to three flood events are sufficient for the crop to mature. The sediments are generated in the upper catchments, through erosion due to the clearing of indigenous trees for agricultural land (Enfors and Gordon, 2007). These deposits are not only beneficial, years of deposits have silted up the two bridges upstream of the spate-irrigation system. Currently, the level of the fields are much higher than the canals, due to these deposits, directing water from the canals towards the fields has become a laborious activity. More labour is required to rebuilt the unlined canals and intake. Diversion bunds are eroded, which need to be rebuilt, and deposits in the canals and under the bridges need to be cleared before the next flood event. Some flood events can be so massive that they destroy crops and infrastructure, as did the floods of December 2003, March 2006 and December 2007. The system can serve a maximum of 300 hectares owned by some 60 households.

Each diversion canal has its own furrow representative, who is appointed by the farmers for an unlimited period. In principle, any member of the furrow can be appointed as representative regardless gender or age, but in practice they are members of the six families who established the furrow system. In case of inadequacy, the farmers can decide to replace the representative, but this rarely happens. At the level of the entire spate-irrigation system, the six representatives form a committee together with the chairperson, secretary and treasurer, who are also appointed for an unlimited period (Komakech et al., submitted). The committee decides upon the allocation of water between the furrows. The representative for each furrow decides on the allocation of the water to the farmers of that furrow. When water is abundant the farmers can be allocated water for six hours whereas at times when the flow is low, a farmer can be given water for 12 hours (to be able to receive sufficient amounts). On Sundays, the water is allocated to the representative of the furrow that receives the allocation as compensation for managing the water. Usually the flow lasts for 4 to 5 days per event, but in good seasons the flow can last for several weeks.

When the flood starts, the farmer who sees the water coming first is allowed to divert the water into his/her farm without approval by the management. This "bonus" system is used as an incentive for people to be vigilant, as they do not have any facility to store the flood flow, and the "use it or loose it" principle applies. Chances of any runoff passing through the river unnoticed are thus minimized. When enough water has entered the farm, the farmer is responsible for closing the intake so that the next farmer can use the water. Box 7.1 gives the rules and regulations of the committees, which have been obtained through interviews with members of the spate-irrigation system.

Box 7.1 Rules and regulations of the Makanya spate-irrigation committees

- To be allocated water in a season, a member must have participated in the preparation of the furrow in that season, which requires a substantial amount of work.

- Any member who has been allocated a piece of land by the village government and whose piece of land is within the command area of any furrow is automatically member of the furrow's committee.

- Any member who fails to take part in the preparation of the furrow is penalised by the other members by taking property from her/his field amounting to the value of the days of work missed. This property is sold to recover the penalty. The penalty agreed by the members is that businessmen pay an equivalent of TSh 6,000 (approx. 4.8US$ equivalent), while non-businessmen pay TSh 2,000 (approx. 1.6US$ equivalent), for one days work. The money obtained from these sales is used to run the system.

- On Sundays the water is allocated to the representative of the furrow receiving the water at that moment.

- If the water in the river is considered enough, water allocation starts from the downstream end.

7.3 Water Sharing between Users of Adjacent Furrows

The Bangalala furrows are described as an example of how water is being allocated between adjacent furrows belonging to the same village, which divert water from the same river. There are five major furrow systems in Bangalala each with a micro dam, of which four divert water from the Vudee River, namely the furrows known as Mghungani, Kinyang'a, Mkanyeni and Manoo. Mghungani and Kinyang'a divert water from Upper-Vudee before the confluence with Ndolwa. In addition, there is the Mondo wa Ng'ombe furrow system which has no micro dam, located after the confluence and upstream of the abstraction points for Mkanyeni and Manoo furrow systems. Mchikatu furrow system abstracts water from a small tributary (Table 7.3; Fig. 7.1). The Mondo wa Ng'ombe furrow system is regarded as part of the upstream zone in the Manoo furro system and shares in the water allocation in this zone. Many farmers in the adjacent furrow systems own plots in both systems, which spreads the risk and may help to avoid conflicts due to the interdependencies (Kemerink et al., submitted). Mchikatu furrow system abstracts water from a small tributary (Table 7.3; Fig. 7.1)

Table 7.3 Characteristics of furrow systems in Bangalala (Vyagusa, 2005).

Name furrow system	Established (Year)	Rehabilitated (Year)	Families served	Command area (ha)	Water supply
Manoo	1936	2001	150	400	Vudee
Mkanyeni	1951	2004	70	40	Vudee
Mondo wa Ng'ombe (no micro dam)	1945	-	40	80	Vudee
Kinyang'a	2000	2004	124	6	Upper-Vudee
Mghungani	1957	2004	115	66	Upper-Vudee
Mchikatu	1959	2000	95	11	Mchikatu

There is an agreement between Manoo and Mkanyeni furrow systems on abstraction practices. Each furrow diverts water for three days in a week from 4 PM until 4 PM the following day after which the irrigation turn is transferred to the other furrow. Once every two weeks, each furrow system has an irrigation turn on Sunday. This agreement is clearly written in the constitution of Mkanyeni furrow system. The constitution of Manoo furrow system refers to this agreement in the following way;

> "Manoo furrow system will co-operate with Mkanyeni furrow system on issues regarding abstraction of water from the river".

There is no agreement between the Manoo and Mkanyeni furrow systems on the main stream and the two upstream furrow systems (Mghungani and Kinyang'a). The upstream furrows abstract water every day even if that means that during dry spells no water is left in the river for the downstream furrows in the same village of Bangalala. One of the reasons is that Kinyang'a and Mghungani abstract water from Upper-Vudee before the confluence with the Ndolwa. The assumption from Kinyang'a and Mghungani water users is that Manoo and Mkanyeni should use the water that comes from the Ndolwa, and not the water that comes from Upper-Vudee. Therefore, the upstream furrow users do not see the need for an agreement. However, the major part of the flow at the diversion point for Manoo and Mkanyeni furrow systems comes from Upper-Vudee even when Kinyang'a and Mghungani abstract water (Mul et al., submitted), consistent with the larger catchment size of the Upper-Vudee. This also seems to be recognised at village level, as a water sharing agreement does exist between Vudee and Bangalala villages as described below.

7.4   WATER ALLOCATION BETWEEN NEIGHBOURING VILLAGES

One would think that, because of the larger distances it would be more difficult to reach agreements between villages than between neighbouring furrows. However, these agreements do exist. Here we review the agreements that emerged between the three villages in the Vudee sub-catchment, Vudee, Ndolwa and Bangalala. The Vudee sub-catchment drains an area of approximately 31.4 km^2. In 2002 the sub-catchment had a population of 9,700 growing at a rate of 1.6 percent per annum (Table 7.4, URT, 2004). There are about 38 furrow systems of which 20 have micro dams. The average size of the micro dams in Vudee and Ndolwa are smaller then the dams in Bangalala. The water available for Bangalala is affected by the two upstream villages. In the highlands, irrigation is mainly used during the dry season and only as supplementary irrigation in the rainy season during dry spells, while in Bangalala supplementary irrigation is almost always needed during the rainy season. Fischer (2008) showed that the occurrence of a critical dry spell in the higlands is less than 10 percent, whereas this is almost 90 percent in Bangalala. The flow in the river is not enough for full irrigation.

Table 7.4 Characteristics of villages using water in Vudee sub-catchment

	Vudee	Ndolwa	Bangalala	Total
Population (source: URT, 2004)	3,800	2,430	3,470	9,700
Catchment area [km^2]	14.2	8.4	8.8	31.4
Large furrow systems [12] [no.]	17	0	1	18
Furrow system including micro dams [no.]	6	9	5	20
Irrigated area in rainy season [ha]	200	30	523[13]	753

7.4.1   Agreement between Ndolwa and Bangalala

Bangalala has been discussing the issue of water sharing with Ndolwa village since the 1940s. In the first agreement of 1949, Ndolwa agreed to release water for downstream uses. In 1958 this was further defined, and, one day per week water would be released for downstream users, and Ndolwa villagers were not allowed to abstract water from the river for any reason on that day. This agreement was respected by all parties until the 1970s when Ndolwa began to experience population and economic growth which, in turn, increased pressure on water resources. The 1974 drought saw Ndolwa abandoning the agreement and stopped releasing water for downstream uses. In 1976 the situation returned to normal, but during dry spells Ndolwa still abstracts water from the river every day without regarding the consequences for farmers in Bangalala.

At present people from Ndolwa village claim that flows from the river and streams do not reach Bangalala village during dry spells even if they would not divert any water. They argue that they have hardly enough water for their own crops during such periods and cannot afford to let water flow downstream. Bangalala villagers are of the opinion that Mkanyeni and

---

[12] Furrow system without micro dam owned by more than one family

[13] Figure refers to irrigated area in Bangalala during good rainy seasons when water is abundant. During dry spells farmers are only allowed to irrigate part of their lands and during the dry season the irrigated area decreases almost to zero. In Vudee and Ndolwa, the irrigated area is almost the same during the rainy season as during the dry season, obtaining on average three harvests per year. This explains why the irrigated area per inhabitant in Vudee and Ndolwa is substantial smaller compared to Bangalala.

Manoo furrow systems mainly depend on water originating from the Ndolwa as the upstream irrigation system in the village abstract water from Upper-Vudee[14]. Both villages appreciate the need to find a solution to the water scarcity in the basin, hence, eight water user groups (see Table 7.5) from both villages formed an association of water user groups in 2004. This association was formed on advice of an NGO called TIP (Traditional Irrigation Improvement Project), which taught them to practice agriculture that conserves water and soil. The association is referred to as "UNYINDO" and the main focus is "to find new water sources" to be shared by farmers in both villages. The potential new water resources could come from the mountain wetlands, which fall within the administrative boundaries of Ndolwa, but in the watershed of the villages on the other side of the mountains. However, since the establishment, no concrete outcomes have been achieved and the UNYINDO only meets irregularly.

Table 7.5 Members of the UNYINDO association of water user groups.

Water User Group	Hamlet	Village	No. families
Heivumba	Masheko	Ndolwa	10
Ndiveni	Ndiveni	Ndolwa	20
Kwanashanja	Mjingo	Ndolwa	40
Kitieni	Kitieni	Ndolwa	40
Kitala	Masheko	Ndolwa	60
Mombo	Mtwana	Ndolwa	60
Mkanyeni	Mkanyeni	Bangalala	70
Manoo	Kwanyungu	Bangalala	150

7.4.2 Agreement between Vudee and Bangalala

Bangalala and Vudee villages have an unwritten agreement on the sharing of water agreed at a meeting between the two villages and is based on historical water sharing arrangements. Vudee farmers are not allowed to irrigate at night. Abstraction is only allowed for the purpose of filling the micro dams. At night, water is left to flow for downstream users in Bangalala village where water is diverted to the furrow system, filling the micro dams, which is used for irrigation the next day. On Sundays, Vudee villagers are not allowed to abstract water from the river so that water can be used by the "environment and animals"[15].

Daily variations are observed in the water levels at the weir downstream of the confluence of the Upper-Vudee and Ndolwa rivers (Fig. 7.4). The fluctuations show a clear diurnal pattern, with the highest flows observed early in the morning, which is consistent with the agreements between Vudee and Bangalala to release water for the downstream village during the night. On Sundays, the decrease of the flow is indeed less than during the other days as indicated in Fig. 7.4, which corresponds with aim for environmental flows on Sundays. Another observation is that there is a stronger drawdown on particular week days. This drawdown is attributed to the abstractions from the Mondo wa Ng'ombe furrow system upstream of the weir, which is diverting water mainly on Wednesdays within the irrigation turns of Manoo furrow system.

The fluctuations were also observed during the dry season, hence, it can conclude from it that the arrangements agreed between Vudee and Bangalala villages are being adhered to even during periods of low flows. It is also observed that the abstractions are about 10 l s^{-1},

[14] However, hydrological data indicates that the irrigation systems of Mkanyeni and Manoo receive substantial part of the water originating from Vudee than from Ndolwa, see also paragraph 7.3.
[15] In reality this water is used by the downstream Bangalala community.

which equals almost 50 percent of the total flow during this period. The fluctuations show the amount of water, which is diverted by furrow systems in Vudee village which adhere to the agreements[16], potentially there are other abstractions which continuously divert the flows, and, therefore, reduce the flow at the weir. Particularly, it is unclear how much is diverted by the people in Ndolwa, who do not have an agreement with the people in Bangalala. During high flows these fluctuations are not observed. This is partly due to the fact that the accuracy of the discharge measurement is reduced with higher flows, and partly because during high flows (in the wet season), the upstream villagers do not need the flow for irrigation.

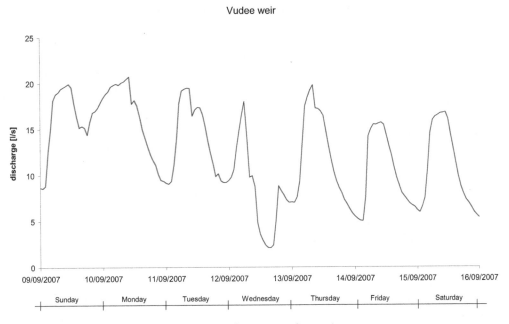

Figure 7.4 Water level fluctuations as a result of upstream abstractions.

The upstream furrow systems development has had a considerable impact on the availability of water in Bangalala. Although the flow at the site of the weir has been perennial as long as people can remember, at some extreme occasions it has been reported to fall dry after a substantial drought period, such as in 1948, 1974, 1977 and in early 2006 (Mul et al., 2006). There does not appear to be a significant increase in the frequency of the river to fall dry. Although the amount of base flow has steadily decreased over the years since the 1950s, which is consistent with the increased activities upstream.

7.5   WATER SHARING BETWEEN DISTANT VILLAGES

No water sharing agreements exist between Bangalala and the village of Makanya, 25 kilometers downstream, where spate-irrigation is practiced. The downstream furrow systems of Bangalala (Manoo and Mkanyeni) attempt to divert all the water from the river. The only

---

[16] In this analysis it is assumed that the furrow systems of Bangalala village upstream of the weir (Kinyang'a and Mghungani) divert continuously water and, therefore, do not contribute to the fluctuations in the discharge at the weir, however, it does impact the total flow at the weir.

flow downstream is leakage from the diversion structure, which not much further completely infiltrates into the sandy river bed. Nowadays, only high intensity rainfall connects the two spatial scales, whereby the flows reach the spate-irrigation system in Makanya for several days.

These flood flows, on which spate-irrigation relies, have increased in occurrence and magnitude over the years, with the most extreme event during the El Niño year (1997), although in 1951 a heavy flood was also observed (Mul et al., 2006). Moreover, water source for the spate-irrigation system in Makanya is not only limited to the Vudee sub-catchment, but in total four sub-catchments in the South Pare Mountains contribute to generating flood flows to the spate-irrigation system. Significant spatial variability of the rainfall in the upstream sub-catchments[17] results in a small number of flash floods of sufficient magnitude in Makanya each year. During the drought of Vuli 2005, it was observed that whereas there was complete crop failure in the upstream catchment, crop yields in the spate-irrigation system near the intakes were not affected as, despite the drought, an extensive flood lasting up to 10 days and two smaller peak events were received in the lowlands.

Water users in the South Pare Mountains have, without the intervention of the national authorities, been able to agree upon sharing the available water. As described in this paper, water sharing agreements were found to exist between irrigators sharing the same furrow, between neighbouring furrows that divert water from the same river, and between neighbouring villages along the same river system. Some agreements date back to the 1940s and they mostly specify water sharing on a rotational basis between irrigators, furrows or villages. However, no water sharing agreements were found at a larger spatial scale beyond the sub-catchment scale (25 km^2), such as between the villages in the Vudee sub-catchment located in the centre of the catchment, and Makanya village.

One may ask the question, why such arrangements between more distant water users did not evolve over time. One line of reasoning is that the hydrological disconnect, that is currently observed in Makanya is the result of increased water abstractions upstream, and in fact the consequence of successful farmer initiatives to harvest more rainfall and to "drought-proof" or "climate-proof" their farming systems. It may be more efficient to use the low flows upstream, where the chances of using the little water productively are higher than losing released water for downstream purposes to transmission losses. In the past, there was no reason for distant villages to develop water sharing agreements, because no upstream-downstream tradeoffs were felt. However, the river that used to be perennial has become ephemeral (Mul et al., 2006). It is important to realise, that not all the water is used upstream, large transmission losses in the sandy river bed captures most of the flow before reaching Makanya. As a result of the increase in demand for a diminishing resource, tradeoffs between upstream and downstream water uses have emerged at an increasingly larger spatial scale.

A contrasting line of reasoning is that the upstream furrow irrigators and the downstream spate irrigators do not compete, since they use water stemming from different parts of the hydrograph: furrow irrigation upstream typically diverts base flows whereas the spate-irrigation downstream depends on flood flows[18] from four different sub-catchments. In the current setting, this would imply that the increasing water consumption in the upstream parts of the catchment does not strongly diminish the availability of water downstream nor

---

[17] Spatial variability in rainfall even occurs in areas with similar altitudes (Mul et al., 2008b).
[18] It must be noted that before the river ceased to be perennial, villagers in Makanya also used the base flow for irrigation.

the frequency of events. Hence, agreements on water sharing between the more distant villages are not required. The same line of reasoning holds to a lesser extent also for the different furrow systems studied, as the farmers in the mountains (such as Vudee and Ndolwa)
mainly irrigate during the dry season, while the farmers downstream in Bangalala use the
water for supplementary irrigation during the rainy season. In this light, it should be noted
that furrow systems in the lower parts developed later than the furrow systems in the upper
parts of the catchment, which could indicate that the downstream farmers adapted their systems in a way that they would make use of the water, which was not already used by the upstream communities. However, due to increased demands for food in the area the competition
over water has increased over time, which potentially hampered such non-competitive water
sharing arrangements, and therefore, the agreements between the neighbouring villages to
share water originating from the same part of the hydrograph may have evolved. This potentially indicates that such agreements are also needed between more distant villages in case
the spate-irrigation system does not longer provide sufficient food to Makanya.

7.6   DISCUSSION AND CONCLUDING REMARKS

Both lines of reasoning indicate that people in the catchment are well aware of the increased
competition over water, and are actively seeking answers, by borrowing from the ways they
have been resolving water allocation problems in the past, namely to build on the social ties
that exist between communities. However, due to the larger spatial scales at which the tradeoffs manifest themselves, the social ties between more distant communities are relatively weak
(Cleaver and Franks, 2005); possibly too weak to build new water sharing arrangements on.
Some would, therefore, argue that it may be necessary to involve more formal levels of government, such as Pangani Basin Water Office, to assist with the establishment of water allocation agreements at the (meso) catchment scale.

Both lines of reasoning also throw up pertinent questions that beg for answers and that re-
quire further research. The argument to use the water as much as possible high up in the
catchment, nearest to where it touched down as rainfall because of efficiency considerations,
raises the question what the impact would be on downstream users, and whether or not
downstream users need to be compensated, and if so, how? Reference is made to the recent
discussions on the payment for environmental services (Hermans and Hellegers, 2005), but
then in reverse mode. It also calls for the establishment of water sharing arrangements at increasing spatial scales. What would be the basis for such arrangements? Why would upstream water users be willing to forego certain immediate benefits? Would the hydrosolidarity concept (Falkenmark and Folke, 2002; Falkenmark and Lundqvist, 1999) be a
mechanism to enhance water sharing practices between the water users at the various levels
in the catchment (Kemerink et al., submitted)? Finally, the argument that the various furrow
systems do not compete over water as they target different parts of the hydrograph needs detailed hydrological evidence: what is the precise impact of increased water consumption of
many small water users scattered all over the upper parts of the catchment, and how would
they impact on the magnitude and frequency of flash floods reaching the catchment's outlet?
What does this imply for the entire Pangani river basin, to which Makanya catchment is hy-
draulically disconnected most of the time? These are some of the research questions that need
to be answered in the near future.

Chapter 8

SYNTHESES AND CONCLUSIONS

The livelihood of the local people in the Makanya catchment is determined for a large part by the occurrence and availability of water, which varies strongly over the catchment. Understanding the hydrological processes in an ungauged catchment requires a multi-method approach, which is described in Chapters 3-6. The human response and adaptation to the local climate and upstream users is presented in Chapter 7.

8.1   FARMING ADAPTATION TO CLIMATOLOGICAL, HYDROLOGICAL AND BIO-PHYSICAL CONSTRAINTS

The long term annual rainfall in the Makanya catchment varies between 800 mm a^{-1} (Tae, at 1,740 m elevation) in the highlands and 500-550 mm a^{-1} in the lowlands (Makanya (650 m) and Same (882 m) data from 1930-2006). The annual rainfall is spread over two growing seasons. The average rainfall amounts to 200 mm season^{-1} in the *Vuli* season and 325 mm season^{-1} in the *Masika* season (Same data, 882 m). These amounts are often insufficient to sustain purely rainfed agriculture. In addition, the inter-seasonal variability is high with standard deviations of 141 mm season^{-1} for *Vuli* and 131 mm season^{-1} for *Masika*. The chance of receiving rainfall of 500 mm season^{-1}, the required amount to grow maize (the preferred crop) is low (Makurira et al., 2007b). On top of that, the intra-seasonal variability is high and the frequency of occurrence of critical dry spells is increasing (Enfors and Gordon, 2007). The probability of critical dry spells within the Makanya catchment depends on the location. Fischer (2008) showed that the occurrence of critical dry spells in the catchment is predominant at lower altitudes. Depending on the water holding capacity of the soil, potential evaporation and the rainfall occurrences, critical dry spells in the valley have a probability of occurrence of more than 90 percent during the growing seasons. At higher altitudes, with better soils and climatologic conditions, critical dry spells have a probability of occurrence of less than 10 percent.

In view of these conditions, it is not strange that settlements first developed in the highlands. Here, optimisation of yields is achieved through supplementary irrigation from the abundant springs. Even during the dry season, people in the highlands are cultivating crops using full irrigation. Increasing population pressure, and the hunt for agricultural lands, changed the predominantly forested hill slopes into agricultural lands (Enfors and Gordon, 2007; IUCN,

2003). The steep topology, in combination with the clearing of forest, accelerated erosion. As the potential for expanding agricultural lands decreased in the highlands, the people moved to the less suitable areas, such as in the Makanya valley. Here, rainfed agriculture requires supplementary irrigation, because of frequent dry spells. Only in the floodplains, where the eroded fertile soils from the highlands have been deposited, the water holding capacity is sufficient to cope with prolonged dry spells (Komakech et al., 2008). In addition, an irrigation network has developed to provide supplementary irrigation to most of the plots near or downstream of a perennial or seasonal river. These irrigation systems are managed by the local community, consisting of the farmers living within the scheme, without any interference from outside (Makurira et al., 2007a). Agreements on sharing water resources between the highlands and the midlands have been in existence for a long time, depending on family relationships.

Only few areas that lack surface water supply have to cope with rainfall alone. An alternative to increasing the water supply to these areas is the use of gully diversions (Makurira et al., 2007b), which collect surface runoff (from gullies, natural catchments or road drains) and which only function directly after an event with high rainfall intensity. In addition, soil and water conservation techniques that improve infiltration and enhance the moisture storage capacity, and hence reduce surface runoff, have a positive effect on the yield. More water is hereby directed into and retained in the field, increasing the capacity of the plants to overcome dry spells. Both techniques also improve the soil structure. Gully diversion adds sediments and nutrients to the field from erosion in the upstream catchment area and soil and water conservation techniques reduce erosion from the plot, retaining the soil.

Finally, farmers in the dry plains near Makanya have resorted to applying spate-irrigation as a way of harvesting the flash floods in absence of perennial flows. The harvested flash floods deposit fertile sediments on the plots and reasonable harvests are generated in an area with limited rainfall. In Same, with similar climatological conditions, farmers do not plant in the *Vuli* season, as rainfall is generally insufficient. However in Makanya, *Vuli* harvests are equally good as *Masika* harvests, as the occasional flash floods generate sufficient buffer in the soil for the plants to grow.

Areas with good local climatological conditions, which are not already used, are becoming increasingly scarce. Less suitable areas are now being used, which pose significant challenges for local people. In absence of reliable surface water supply, soil and water management techniques are essential in these areas for generating good yields. Many things can be learned from the way local people have adapted to the different local climates in the Makanya catchment. This local knowledge can also be used for developing strategies to cope with future climate change scenarios.

8.2 Hydrological Process Understanding and Modelling

Over 80 percent of the population of the South Pare Mountains lives on subsistence farming, while rainfed agriculture in many parts of the catchment is unsustainable. The majority of the people, therefore, rely on the availability of water in the streams and in the groundwater for supplementary irrigation to sustain their livelihoods. Understanding the occurrences and prevalence of these water resources, how they respond to rainfall and droughts, is important in order to manage and utilise the water resources.

Irrigation activities involve supplementary irrigation during dry spells and full irrigation during the dry season. Few storage structures exist, which often only act as overnight storage. Irrigation therefore mainly depends on base flow. The spring analyses show that there is an intensive spring network in the area, defined by the geology. Most springs originate in the highlands, which discharge the largest amount of water. Springs on the eastern side of the Pare Mountains have considerably higher base flow. The main dipping direction of the geological formations is towards the eastern side of the South Pare Mountains. Some water flows through fissures and faults, along the ridges, to the eastern, Mbaga catchment, sustaining this larger base flow. Smaller springs do occur in the valley, but their discharge is low compared to the springs in the highlands. In addition, the water quality of these lowland springs is poor, with some localised springs containing high concentrations of fluoride and sulphate. Hence, the hydrological processes sustaining these highland springs are merely enough to support the livelihoods of the local people. Water quality samples taken before and after the rainy season have similar mineral compositions and concentrations, indicating that the source of the water comes from a large system with a relatively long residence time. This indicates that unlike the rainfall, the springs are a reliable source of water for the farmers.

During rainfall events, most of the runoff comes from groundwater. During small events this amounts to more than 95 percent of the total flow. During an observed extreme event, the groundwater contribution was close to 50 percent, increasing the groundwater flow from less than 10 l s^{-1} to at least 2 m^3 s^{-1}, within 1 hour. It is quite remarkable that the groundwater system responds so quickly to the rainfall events and with such a sharp rise. It is hypothesised that displacement of groundwater is responsible for the rapid increase of groundwater flow (Beven, 2004; Petry et al., 2002; Uhlenbrook and Wenninger, 2006). This indicates, that the groundwater system, supplying the streams have a large extent, as it is responsible for this rapid and substantial increase. Direct runoff happens only during extreme events and even then, only during the high intensity periods. Although this does not mean that there is absolutely no surface runoff, it generally re-infiltrates before it reaches a stream. During these events, the local aquifers are replenished. These aquifers sustain base flow during dry spells and the dry season, which the farmers rely on.

In the valley, water coming from the mountains travels through the riverbed, recharging the local aquifer along the way. The river travels a certain distance through the valley, depending on the infiltration rate of the alluvium and the discharge coming from the mountains. Floods recharge the alluvium and sometimes reach the outlet of the catchment near Makanya. At several locations in the catchment alluvial fans are observed, where the river infiltrates over a wider area. In addition, extreme events, overtopping the natural bunds, recharge aquifers beneath local floodplains, depositing fertile sediments in the process. In these areas, water is readily available, although the crops are at high risk of damage from flooding. However, the benefit of farming in this area outweighs the risks.

The conceptual model developed confirms the hypothesised groundwater flow path. A considerable amount of runoff discharges towards the neighbouring catchment through the underground. When this flow path is saturated, water is reconnected with the stream flow towards the Makanya catchment and produces an increase of the flow at high water stages. The model has shown to be able to reproduce the flows at the foot of the mountain reasonably well, even on high intensity data. However, to be able to model the flows at the outlet of the catchment, additional processes need to be incorporated, such as the processes occurring in the alluvial fan. There is currently a lack of knowledge, thus a further study into the hydrological processes of these alluvial fans is necessary. This will also provide insight into the impact of upstream water use on the water availability in Makanya, which currently can only be qualitatively determined.

8.3   IMPACT OF FARMING ACTIVITIES ON HYDROLOGY

Agricultural developments in this area, particularly with regard to the water use for irrigation, have affected the water availability in other parts of the catchment. Increased water use, through irrigation in the highlands and midlands, reduced flows into the valley. Water sharing arrangements between the highland and midland villages have developed over time as water use increased. The reduction, over the years, has gotten to an extent that the flow into the alluvium does not exceed the infiltration capacity of the river bed during base flow. As a result, the perennial river at the outlet of the catchment has changed into an intermittent river. Where the perennial river used to discharge into a wetland, the wetland has since many years dried up. Groundwater stored in the alluvium is used by the Sisal Estate situated near the outlet of the catchment. The wetlands located on the alluvial fan near Makanya have been transformed into a spate-irrigation system, retaining the flash floods and the soils deposited from the eroded highlands.

Next to the water use in the highlands, conversion from forested land to agricultural lands has had an impact on the water availability (IUCN (2003) estimated that most of the highland area was covered in forest before it became inhabited). Over the last 50 years, large scale conversion from forest to agricultural land occurred (Enfors and Gordon, 2007). The existing runoff time series in the Makanya catchment are not long enough to observe these changes. However, the extent of the land use change is of such a magnitude that it must have made a substantial difference. Steep slopes are being cultivated without any means of protection, such as terraces and contour ridges. Erosion from the plots is deposited in the alluvial fans and floodplains. This can be seen clearly at the road and railway bridges near Makanya, where the flow-through area is substantially reduced by sediment deposits, and where the bridges overtop regularly, as a result.

In the Makanya catchment, changes in the runoff regime have been observed at several locations. Base flows have decreased, whereas at some locations the streams dried out completely. Wetlands have dried up as a result of the upstream water use. In addition, peak flows, over-topping the main road between Dar Es Salaam and Arusha, have occurred several times over the last five years. Because of the lack of historical data, it was not possible to determine whether the frequencies and magnitudes of the floods have actually increased.

The use of soil and water management techniques in rainfed agriculture may have similar impacts on downstream water availability. As an example, gully diversion captures water during flash floods generated in an upstream catchment area. As the diversion structure diverts water into the fields, the structure reduces the floodpeak. Currently there is no base flow is this area of the catchment, and therefore the impact on low flows in unclear. Replenishment of the groundwater through these diversions could improve the water quantity and quality of the groundwater, making it a more reliable resource for the local people. In addition, small floods generally do not even reach the spate-irrigation system, with or without the diversions. Instead of infiltrating into the alluvium, the water is now used productively. For large floods, the amount a gully diversion can divert is small, and diversons would not deprive down-stream users of water, but may reduce devastating floodpeaks. Upscaling this technique within the catchment may have a considerable impact. Many diversions could result in that the discharge no longer exceeding the infiltration capacity of the alluvium so that flash floods no longer reach the spate-irrigation near the outlet.

Currently, the catchment is in a fragile balance. Water is used by most farmers in all areas in the catchment, and yields are produced by each sub-system. However, increasing population densities and the absence of a catchment-wide organisation, may result in an unequal sharing of the catchment's water resources.

REFERENCES

Acrement, G.J. and Schneider, V.R., 1990. Guide for selection Manning's roughness coefficients for natural channels and floodplains, United States Geological Survey Water-supply paper 2339.

Adams, W.M., Potkanski, T. and Sutton, J.E.G., 1994. Indigenous farmer-managed irrigation in Sonjo, Tanzania. The Geographical Journal, 160(1): 17-32.

Allen, R.G., Pereira, L.S., Raes, D. and Smith, M., 1998. Crop Evapotranspiration: Guidelines for computing crop water requirements, FAO, Rome.

Appelo, C.A.J. and Postma, D., 1993. Geochemistry, groundwater and pollution. Balkema, Rotterdam, The Netherlands.

Bagnall, P.S., 1963. The geology of the North Pare Mountains. Bulletin of the Geological Survey of Tangayika, 10: 7-16.

Barham, E., 2001. Ecological boundaries as community boundaries: the politics of watersheds. Society and Natural Resources 14: 181-191.

Belmonte, A.C. and Beltran, F.S., 2001. Flood events in Mediterranean ephemeral streams (ramblas) in Valencia region, Spain. Catena, 45: 229-249.

Beven, K., 1993. Prophesy, reality and uncertainty in distributed hydrological modelling. Advances in Water Resources, 16: 41-51.

Beven, K., 2004. Robert E. Horton and abrupt rises of groundwater. Hydrological Processes, 18(18): 3687-3696.

Beven, K.J. and Binley, A.M., 1992. The future of distributed models: model calibration and uncertainty prediction. Hydrological Processes, 6: 279-298.

Bhatt, Y. et al., 2006. Smallholder system innovations in integrated watershed management (SSI): Strategies of water for food and environmental security in drought-prone tropical and subtropical agro-ecosystems, IWMI Colombo, Sri Lanka.

Binley, A. and Kemna, A., 2005. DC resistivity and induced polarization methods. In: Y. Rubin, Hubbard, S.S. (Editor), Hydrogeophysics. Springer, pp. 129-156.

Blomquist, W. and Schlager, E., 2005. Political pitfalls of integrated watershed management. Society and Natural Resources 18: 101-117.

Blume, T., Zehe, E., Reusser, D.E., Iroume, A. and Bronstert, A., 2008. Investigation of runoff generation in a pristine, poorly gauged catchment in the Chilean Andes I: A multi-method experimental study. Hydrological Processes: DOI: 10.1002/hyp.6971.

Bond, B.J. et al., 2002. The zone of vegetation influence on baseflow revealed by diel patterns of streamflow and vegetation water use in a headwater basin. Hydrological Processes, 16: 1671-1677.

Bormann, H., Breuer, L., Gräff, T. and Huisman, J.A., 2007. Analysing the effect of soil properties changes associated with land use changes on the simulated water balance: A comparison of three hydrological catchment for scenario analysis. Ecological Modelling, 209: 29-40.

Bosch, J.M. and Hewlett, J.D., 1982. A review of catchment experiments to determine the effects of vegetation changes on water yield and evapotranspiration. Journal of Hydrology, 55: 3-23.

Bren, L.J., 1997. Effects of slope vegetation removal on the diurnal variations of a small mountain stream. Water Resources Research, 33(2): 321-331.

Bronson, K.F. et al., 2004. Carbon and nitrogen pools of southern plains cropland and grassland soils. Soil Science Society American Journal 68: 1695-1704.

Brown, A.E., Zhang, L., McMahon, T.A., Western, A.W. and Vertessy, R.A., 2005. A review of paired catchment studies for determining changes in water yield resulting from alteration in vegetation. Journal of Hydrology, 310: 28-61.

Brown, V.A., McDonnell, J.J., Burns, D.A. and Kendall, C., 1999. The role of event water, a rapid shallow flow component, and catchment size in summer stormflow. Journal of Hydrology, 217: 171-190.

Bruijnzeel, L.A., 1988. (De)forestation and dry season flow in the tropics: a closer look. Journal of Tropical Forest Science, 1(3): 229-243.

Burt, T.P., 1979. Diurnal variations in stream discharge and throughflow during a period of low flow. Journal of Hydrology, 41: 291-301.

Butler, J.J., Jr. et al., 2007. A field investigation of phreatophyte-induced fluctuations in the water table. Water Resources Research, 43.

Buttle, J.M., 1994. Isotope hydrograph separations and rapid delivery of pre-event water from drainage basins. Progress in Physical Geography 18: 16–41.

Caissie, D., Pollock, T.L. and Cunjak, R.A., 1996. Variation in stream water chemistry and hydrograph separation in a small drainage basin. Journal of Hydrology, 178: 137-157.

Calder, I.R., 1999. The Blue Revolution. Earthscan Publications, London, 192 pp.

Calder, I.R. et al., 1995. The impact of land use change on water resources in sub-Saharan Africa: a modelling study of Lake Malawi. Journal of Hydrology, 170: 123-135.

Chow, V.T., 1959. Open-channel hydraulics. McGraw-Hill Book Company.

Cleaver, F. and Franks, T., 2005. How institutions elude design: river basin management and sustainable livelihoods. 12, Bradford Centre for International Development. University of Bradford, Bradford.

De Groen, M.M., 2002. Modelling interception and transpiration at monthly time steps; introducing daily variability through Markov chains. PhD Thesis, IHE-Delft, Delft, the Netherlands.

Didszun, J. and Uhlenbrook, S., 2008. Scaling of dominant runoff generation processes: Nested catchments approach using multiple tracers. Water Resources Research, 44.

Dye, P.J. and Poulter, A.G., 1995. A field demonstration of the effect on streamflow of clearing invasive pine and wattle trees from a riparian zone. Suid-Afrikaanse Bosboutyskrift, 173: 27-30.

Edwards, K.A., 1979. The water balance of the Mbeya experimental catchments. East African Agricultural and Forestry Journal, 43: 231-247.

Enfors, E. and Gordon, L., 2007. Analyzing resilience in dryland agro-ecosystems. A case study of the Makanya catchment in Tanzania over the past 50 years. Land degradation and development 18: 680-696.

Falkenmark, M., 1995. Coping with Water Scarcity under Rapid Population Growth, Conference of SADC Ministers, Pretoria.

Falkenmark, M. and Folke, C., 2002. The ethics of socio-ecohydrological catchment management: towards hydrosolidarity. Hydrology and Earth System Sciences 6(1): 1-9.

Falkenmark, M. and Lundqvist, J., 1999. Towards upstream/downstream hydrosolidarity; introduction, SIWI/IWRA Seminar. Towards upstream/downstream hydrosolidarity, Stockholm, pp. 11-15.

Fenicia, F., Savenije, H.H.G. and Avdeeva, Y., 2008. Anomaly in the rainfall-runoff behaviour of the Meuse catchment. Climate, land use, or land use management? Hydrology and Earth System Sciences Discussion, 5: 1787-1819.

Fenicia, F., Savenije, H.H.G., Matgen, P. and Pfister, L., 2006. Is the groundwater reservoir linear? Learning from data in hydrological modelling. Hydrology and Earth System Sciences, 10: 139-150.

Fenicia, F., Savenije, H.H.G., Matgen, P. and Pfister, L., 2007. A comparison of alternative multiobjective calibration strategies for hydrological modelling. Water Resources Research, 43.

Fischer, B.M.C., 2008. Spatial variability of dry spells, a spatial and temporal rainfall analysis of the Pangani Basin & Makanya catchment, Tanzania. MSc Thesis, Delft University of Technology, Delft.

Fleuret, P., 1985. The social organization of water control in the Taita Hills, Kenya. American Ethnologist., 12(1): 103-118.

Foody, G.M., Ghoneim, E.M. and Arnell, N.W., 2004. Predicting locations sensitive to flash flooding in an arid environment. Journal of Hydrology, 292: 48-58.

Franzluebbers, A.J., Stuedemann, J.A., Schomberg, H.H. and Wilkinson, S.R., 2000. Soil organic C and N pools under long-term pasture management in the Southern Piedmont USA. Soil Biology Biochemistry 32: 469-478.

Frederickson, G.C. and Criss, R.E., 1999. Isotope hydrologic and residence times of the unimpounded Meramec River Basin, Missouri. Chemical Geology, 157: 303-317.

French, R.H., 1986. Open-channel hydraulics. McGraw-Hill Book Company, 705 pp.

Gaume, E., Livet, M. and Desbordes, M., 2003. Study of the hydrological processes during the Avene river extraordinary flood (South of France): 6-7 October 1997. Physics and Chemistry of the Earth, 28: 263-267.

Gaume, E., Livet, M., Desbordes, M. and Villeneuve, J.-P., 2004. Hydrological analysis of the river Aude, France, flash flood on 12 and 13 November 1999. Journal of Hydrology 286: 135-154.

Grove, A., 1993. Water use by the Chagga on Kilimanjaro. African Affairs 92: 431-448.

Guzman, J.A. and Chu, M.L., 2003. SPELL-Stat statistical analysis program, Universidad Industrial de Santander, Colombia.

Hermans, L. and Hellegers, P., 2005. A "new economy" for water for food and ecosystems; synthesis report of E-forum results, FAO/Netherlands International Conference Water for Food and Ecosystems, The Hague

Hoekstra, A.Y. and Chapagain, A.K., 2007. Water foodprints of nations: water use by people as a function of their consumption pattern. Water Resources Management, 21(1): 35-48.

Hooper, R.P. and Shoemaker, C.A., 1986. A comparison of chemical and isotopic hydrograph separation. Water Resources Research, 22: 1444-1454.

Hudson, N.W., 1993. Field measurement of soil erosion and runoff, FAO, Rome.

IUCN, 2003. The Pangani River Basin: A Situation Analysis, IUCN Eastern Africa Regional Office, Nairobi.

Jewitt, G.P.W. and Görgens, A.H.M., 2000. Scale and model Interfaces in the context of integrated water resources management for the rivers of Kruger national park, Water Research Commission.

Jothityangkoon, C., Sivapalan, M. and Farmer, D.L., 2001. Process controls of water balance variability in a large semi-arid catchment: downward approach to hydrological model development. Journal of Hydrology, 254: 174-198.

Kebede, S., Travi, Y., Alemayehu, T. and Ayenew, T., 2005. Groundwater recharge, circulation and geochemical evolution in the source region of the Blue Nile River, Ethiopia. Applied Geochemistry, 20: 1658– 1676.

Kemerink, J.S., Ahlers, R. and Zaag, P.v.d., submitted. Driving forces behind the development of water sharing arrangements in traditional irrigation systems in northern Tanzania. Society and Natural Resources.

Kessler, T., Mul, M.L., Bohté, R., Uhlenbrook, S. and Savenije, H.H.G., 2008. Investigation of runoff generation responses in steep, semi-arid headwater catchments, South Pare Mountains, Tanzania, WaterNet/WARFSA/GWP-SA annual symposium 29-31 Oct 2008, Johannesburg, South Africa.

Klemeš, V., 1983. Conceptualisation and scale in hydrology. Journal of Hydrology, 65: 1-23.

Kobayashi, D., Suzuki, K. and Nomura, M., 1990. Diurnal fluctuations in streamflow and specific conductance during drought periods. Journal of Hydrology, 115: 105-114.

Koch, K., Wenninger, J., Uhlenbrook, S. and Bonnell, M., 2008. Electrical resistivity tomography (ERT) for identifying hillslope processes in the Black Forest Mountains, Germany. Hydrological Processes.

Komakech, C.H., Mul, M.L., van der Zaag, P. and Rwehumbiza, F.B.R., submitted. Potential of spate irrigation in improving rural livelihoods: case of Makanya spate irrigation system, Tanzania. Agricultural Water Management.

Komakech, C.H., Van der Zaag, P., Jonoski, A. and Van Koppen, B., 2008. Can actor network theory help to understand water sharing practices in Makanya catchment, Tanzania?, WaterNet/WARFSA/GWP-SA annual symposium, Lusaka, Zambia.

Kosgei, J.R., Jewitt, G.P.W., Kongo, V.M. and Lorentz, S.A., 2007. The influence of tillage on field scale water fluxes and maize yields in semi-arid environments: A case study of Potshini catchment, South Africa. Physics and Chemistry of the Earth, 32(15-18): 1117-1126.

Kuczera, G. and Mroczkowski, M., 1998. Assessment of hydrologic parameter uncertainty and the worth of multiresponse data. Water Resources Research, 34(6): 1481-1489.

Laudon, H. and Slaymaker, O., 1997. Hydrograph separation using stable isotopes, silica and electrical conductivity: an alpine example. Journal of Hydrology, 201: 82-101.

Liu, J. and Savenije, H.H.G., 2008. Time to break the silence around virtual-water imports. Nature, 453: 587.

Loheide, S.P.I., Butler, J.J.J. and Gorelick, S.M., 2005. Estimation of groundwater consumption by phreatophytes using diurnal water table fluctuations: A saturated-unsaturated flow assessment. Water Resources Research, 41: W07030.

Loke, H.M., 2003. 2-D and 3-D Electrical Imaging Surveys.

Lørup, J.K., Refsgaard, J.C. and Mazvimavi, D., 1998. Assessing the effect of land use change on catchment runoff by combined use of statistical tests and hydrological modelling: Case studies from Zimbabwe. Journal of Hydrology, 205(3-4): 147-163.

Makurira, H., Mul, M.L., Vyagusa, N.F., Uhlenbrook, S. and Savenije, H.H.G., 2007a. Evaluation of community-driven smallholder irrigation in dryland South Pare Mountains, Tanzania: A case study of Manoo micro dam. Physics and Chemistry of the Earth 32(15-18): 1090-1097.

Makurira, H., Savenije, H.H.G. and Uhlenbrook, S., 2007b. Towards a better understanding of water partitioning processes for improved smallholder rainfed agricultural systems: A case study of Makanya catchment, Tanzania. Physics and Chemistry of the Earth, 32(15-18): 1082-1089.

Makurira, H., Savenije, H.H.G. and Uhlenbrook, S., 2008a. Modelling field scale water partitioning processes using on-site observed data in rainfed agricultural systems: A case study of Makanya Catchment in South Pare Mountains, Tanzania, WaterNet/WARFSA/GWP-SA annual symposium 29-31 Oct 2008, Johannesburg, South Africa.

Makurira, H., Savenije, H.H.G., Uhlenbrook, S., Rockström, J. and Senzanje, A., 2008b. Investigating the water balance of on-farm techniques for improved crop productivity in rainfed systems: A case study of Makanya catchment, Tanzania. Physics and Chemistry of the Earth.

Manzungu, E., 2004. Water for All: Improving water resource governance in Southern Africa, International Institute for Environment and Development IIED, London.

Matsubayashi, U., Velasquez, G.T. and Takagi, F., 1993. Hydrograph separation and flow analysis by specific electrical conductance of water. Journal of Hydrology 152(1–4): 179–199.

Mazor, E., 1991. Chemical and isotopic groundwater hydrology. Open University press, UK.

Mazvimavi, D., 2003. Estimation of flow characteristics of ungauged catchments, case study in Zimbabwe. PhD Thesis, Wageningen University.

McGlynn, B., McDonnell, J.J., Stewart, M. and Seibert, J., 2003. On the relationships between catchment scale and streamwater mean residence time. Hydrological Processes, 17: 175-181.

Mehari, A., van Koppen, B., McCartney, M. and Lankford, B., 2008. Unchartered innovation? Local reforms of national formal water management in the Mkoji sub-catchment, Tanzania. Physics and Chemistry of the Earth.

Mehari, A., van Steenbergen, F. and Schultz, B., 2005. Water rights and rules, and management in spate irrigation system, African Water Laws: Plural Legislative Frameworks for Rural Water Management in Africa, Gauteng, South Africa.

Monteith, J.L., 1965. Evaporation and environment. Symp. Soc. Exp. Biol., 19: 205–234.

Mroczkowski, M., Raper, G.P. and Kuczera, G., 1997. The quest for more powerful validation of conceptual catchment models. Water Resources Research, 33(10): 2325-2335.

Muhongo, S. and Lenoir, J.L., 1994. Pan-African granulite-facies metamorphism in the Mozambique belt of Tanzania: U-Pb zircon geochronology. Journal of the Geological Society, London 151: 343-347.

Mul, M.L. et al., submitted. Water allocation practices among smallholder farmers in South Pare Mountains, Tanzania; The issue of scales. Agricultural Water Management.

Mul, M.L., Mutiibwa, R.K., Foppen, J.W.A., Uhlenbrook, S. and Savenije, H.H.G., 2007a. Identification of groundwater flow systems using geological mapping and chemical spring analysis in South Pare Mountains, Tanzania. Physics and Chemistry of the Earth, 32(15-18): 1015-1022.

Mul, M.L., Mutiibwa, R.K., Uhlenbrook, S. and Savenije, H.H.G., 2008a. Hydrograph separation using hydrochemical tracers in the Makanya catchment, Tanzania. Physics and Chemistry of the Earth, 33(1-2): 151-156.

Mul, M.L., Savenije, H.H.G. and Uhlenbrook, S., 2007b. Base flow fluctuations in a forested hill slope, northern Tanzania, WaterNet/WARFSA/GWP-SA annual symposium, Lusaka, Zambia.

Mul, M.L., Savenije, H.H.G. and Uhlenbrook, S., 2008b. Spatial rainfall variability and runoff response during an extreme event in a semi-arid catchment in the South Pare Mountains, Tanzania. Hydrological Earth System Sciences Discussion.

Mul, M.L., Savenije, H.H.G. and Uhlenbrook, S., 2008c. Understanding key hydrological processes in a small semi-arid catchment in Tanzania by using a multi-method approach, WaterNet/WARFSA/GWP-SA annual symposium, Johannesburg, South Africa.

Mul, M.L., Savenije, H.H.G., Uhlenbrook, S. and Voogt, M.P., 2006. Hydrological assessment of Makanya catchment in South Pare Mountains, semiarid northern Tanzania, Fifth FRIEND World Conference -Climate Variability and Change—Hydrological Impacts-. IAHS, Havana, Cuba, pp. 37–43.

Murty, D., Kirschenbaum, M.U.F., McMurtie, R.E. and McGilvray, H., 2002. Does conversion of forest to agricultural land change soil carbon and nitrogen? A review of the literature. Global Change Biology 8: 105-123.

Mutabazi, K.D. et al., 2005. Economic viability of RWH in semi-arid areas: policy reflections on transformation of dryland agriculture, 7th annual general meeting of Agricultural Economics Society of Tanzania (AGREST), Dodoma, Tanzania.

Mutakyahwa, M.K.D., Ikingura, J.R. and Mruma, A.H., 2001. Geology and geochemistry of Bauxite deposits in Lushoto District, Usambara Mountains, Tanzania. Journal of African Earth Sciences, 36: 357-369.

Mutiibwa, R.K., 2006. Groundwater flow systems in the Makanya catchment of Tanzania. MSc Thesis, UNESCO-IHE, Delft.

Mwamfupe, D.G., 2002. The role of non-farm activities in household economies: a case study of Pangani Basin, Tanzania. In: J.O. Ngana (Editor), Water resources management: the case of Pangani River Basin. Issues and approaches. Dar es Salaam University Press, Dar es Salaam, pp. 39 – 47.

Mwenge Kahinda, J.-M., Rockström, J., Taigbenu, A.E. and Dimes, J., 2007. Rainwater harvesting to enhance water productivity of rainfed agriculture in the semi-arid Zimbabwe. Physics and Chemistry of the Earth, 32(15-18): 1068-1073.

Nash, J.E. and Sutcliffe, J.V., 1970. River flow forecasting through conceptual models. Part 1 - a discussion of principles. Journal of Hydrology, 10: 282-290.

Newmark, W.D., 1998. Forest area, fragmentation, and loss in the Eastern Arc Mountains: implication for the conservation of biological diversity. Journal of East African Natural History, 87: 1-8.

Ngigi, S.N., Savenije, H.H.G., Thome, J.N., Rockström, J. and Penning de Vries, F.W.T., 2005. Agro-hydrological evaluation of on-farm rainwater storage systems for supplementary irrigation in Laikipia district, Kenya. Agricultural Water Management, 73(1): 21-41.

Norbert, J., Moges, S.A. and Kachroo, R.K., 2002. Assessment of mapping of sustainability of rainfed agriculture in Pangani basin using dry spell analysis. In: J.O. Ngana (Editor), Water resources management , the case of the Pangani river basin, Issues and approaches. Dar Es Salaam University Press, Dar Es Salaam, Tanzania.

Nyagwambo, N.L., 2006. Groundwater recharge estimation and water resources assessment in a tropical crystalline basement aquifer. PhD Thesis, UNESCO-IHE, Delft, 170 pp.

Ott, B. and Uhlenbrook, S., 2004. Quantifying the impact of land use changes at the event and seasonal time scale using a process-oriented catchment model. Hydrology and Earth System Sciences, 8(1): 62-78.

Parkhurst, D.L. and Appelo, C.A.J., 1999. User's guide to PHREEQC (version 2) – a computer programme for speciation, batch-reaction, one-dimensional transport, and inverse geochemical calculations, US Geological Survey, US Department of the Interior, Denver, Colorado.

PBWO, 2006. Pangani basin flow assessment initiative. Hydrology and system analysis. Volume 1 of 2: The hydrology of the Pangani river basin, PBWO, Moshi.

Penman, 1948. Natural evaporation from open water, bare soil and grass. Proceedings of the Royal Society of London, 193: 120-145.

Petry, J., Soulsby, C., Malcolm, I.A. and Youngson, A.F., 2002. Hydrological controls on nutrient concentrations and fluxes in agricultural catchments. The Science of the Total Environment, 294: 120-145.

Pinder, G.F. and Jones, J.F., 1969. Determination of the ground-water component of peak discharge from the chemistry of total runoff. Water Resources Research, 5: 438-445.

Plummer, L.N., Busby, J.F., Lee, R.W. and Hanshaw, B.B., 1990. Geochemical modelling in the Madison aquifer in parts of Montana, Wyoming and South Dakota. Water Resources Research, 26: 1981–2014.

Potkanski, T. and Adams, W.M., 1998. Water scarcity, property regimes and irrigation management in Sonjo, Tanzania. Journal of Development Studies, 34(4): 86-116.

Prinsloo, F.W. and Scott, D.F., 1999. Streamflow responses to the clearing of alien invasive trees from riparian zones at three sites in the Western Cape Province. South African Forestry Journal, 185: 1-7.

Reggiani, P., Sivapalan, M. and Hassanizadeh, S.M., 1998. A unifying framework for watershed thermodynamics: balance equations for mass, momentum, energy, entropy and the 2nd law of thermodynamics. Advances in Water Resources, 22(4): 367-398.

Rico, M., Benito, G. and Barnolas, A., 2001. Combined Palaeoflood and rainfall assessment of mountain floods (Spanish Pyrenees). Journal of Hydrology, 245: 59-72.

Rockström, J., 2000. Water resources management in smallholder farms in eastern and southern Africa: An overview. Physics and Chemistry of the Earth, 25(3): 275-283.

Rockström, J. et al., 2004. A watershed approach to upgrade rainfed agriculture in water scarce regions through water systems innovations: an integrated research initiative on water for food and rural livelihoods in balance with ecosystem functions. Physics and Chemistry of the Earth, 29: 1109-1118.

Sahin, V. and Hall, M.J., 1996. The effects of afforestation and deforestation on water yields. Journal of Hydrology, 178(1/4): 293-309.

Sandström, K., 1995. Forest and water – friends or foes. Hydrological implications of deforestation and land degradation in semi-arid Tanzania. PhD Thesis, University of Linköping, Linköping.

Savenije, H.H.G., 1997. Determination of evaporation from a catchment water balance at a monthly time scale. hydrological Earth System Sciences, 1: 93-100.

Savenije, H.H.G., 2000. Water scarcity indicators; the deception of the numbers. Physics and Chemistry of the Earth, 25(3): 199-204.

Savenije, H.H.G., 2001. Equifinality, a blessings in disguise? Hydrological processes, 15(14): 2835-2838.

Savenije, H.H.G., 2004. The importance of interception and why we should delete the term evapotranspiration from our vocabulary. Hydrological Processes, 18(8): 1507-1511.

Scott, D.F., Prinsloo, F.W., Moses, G., Mehlomakulu, M. and Simmers, A.D.A., 2000. A re-analysis of the South African afforestation experimental data, South Africa

Shanley, J.B., Kendall, C., Smith, T.E., Wolock, D.M. and McDonnell, J.J., 2002. Controls on old and new water contributions to stream flow at some nested catchments in Vermont, USA. Hydrological Processes, 16: 589-609.

Shuttleworth, W.J., 1993. Evaporation. In: D.R. Maidment (Editor), Handbook of Hydrology. McGraw-Hill, New York, pp. 4.1-4.53.

Sivapalan, M., 2003. Process complexity at hillslope scale, process simplicity at the watershed scale: is there a connection? Hydrological Processes, 17: 1037-1041.

Sivapalan, M. et al., 2003. IAHS Decade on predictions in ungauged basins (PUB), 2003-2012: Shaping an exciting future for the hydrological sciences. Hydrological Sciences Journal, 48(6): 857-880.

Sklash, M.G. and Farvolden, R.N., 1979. The role of groundwater in storm runoff. Journal of Hydrology, 43: 45-65.

Sokile, C.S., Kashaigili, J.J. and Kadigi, R.M.J., 2003. Towards an integrated water resource management in Tanzania: the role of appropriate institutional framework in Rufiji Basin. Physics and Chemistry of the Earth, 28(20-27): 1015-1023.

Sokile, C.S. and van Koppen, B., 2004. Local water rights and local water user entities: the unsung heroines of water resource management in Tanzania. Physics and Chemistry of the Earth, 29: 1349-1356.

Solomatine, D.P. and Dibike, Y.B., 1999. Automatic calibration of groundwater models using global optimization techniques. Hydrological Sciences Journal, 44(6): 879-894.

Stuyfzand, P.J., 1998. Patterns in groundwater chemistry resulting from Groundwater Flow. Hydrogeology Journal 7: 15–26.

SUA, 2003. Baseline report: Pangani Basin, Smallholder system innovations in integrated watershed management (SSI), 1st progress report.

Swallow, B., Johnson, N., Meinzen-Dick, R. and Knox, A., 2006. The challenges of inclusive cross-scale collective action in watersheds. Water International 31(3): 361-376.

Swallow, B.M., Garrity, D.P. and van Noordwijk, M., 2001. The effects of scales, flows and filters on property rights and collective action in watershed management. Water Policy, 3: 457-474.

Swatuk, L.A., 2005. challenges to implementing IWRM in Southern Africa. Physics and Chemistry of the Earth, 30: 872-880.

Tardy, Y., 1971. Characterization of the principal weathering types by the geochemistry of water from some European and African crystalline massifs. Chemical Geology, 7: 253–271.

Tardy, Y., Bustillo, V. and Boeglin, J.-L., 2004. Geochemistry applied to the watershed survey: hydrograph separation, erosion and soil dynamics. A case study: the basin of the Niger River, Africa. Applied Geochemistry, 19: 469-518.

Tetzlaff, D., Waldron, S., Brewer, M.J. and Soulsby, C., 2007. Assessing nested hydrological and hydrochemical behaviour of a mesoscale catchment using continuous tracer data. Journal of Hydrology, 336: 430-443.

Tiffen, M., Mortimore, M. and Gichuki, F., 1994. More people, less erosion. environmental recovery in Kenya. Wiley, Chichester.

TIP, 2004. Farmers' innovations in traditional irrigation improvement. Pipe conveyance system at Kwa Mlombola water user group, Ndambwe village, Mwanga district.

Turpie, J., Ngaga, Y. and Karanja, F., 2003. A preliminary economic assessment of water resources of the Pangani river basin, Tanzania: economic value, incentives for sustainable use and mechanisms for financing management., PBWO, Moshi.

Uhlenbrook, S., Didszun, J. and Wenninger, J., 2008. Source areas and mixing of runoff components at the hillslope scale - a multi-method technical approach. Hydrological Sciences Journal, 53(3).

Uhlenbrook, S., Frey, M., Leibundgut, C. and Maloszewski, P., 2002. Hydrograph separations in a mesoscale mountainous basin at event and seasonal timescales. Water Resources Research, 38(6): 1-13.

Uhlenbrook, S. and Hoeg, S., 2003. Quantifying uncertainties in tracer-based hydrograph separations: a case study for two-, three- and five-component hydrograph separations in a mountainous catchment. Hydrological Processes 17(2): 431-453.

Uhlenbrook, S., McDonnell, J. and Leibundgut, C., 2001. Foreword to the special issue: Runoff generation and implication for the river basin modelling. Freiburger Schriften zur Hydrologie, 13: 4-13.

Uhlenbrook, S., Roser, S. and Tilch, N., 2004. Hydrological process representation at the meso-scale: the potential of a distributed, conceptual catchment model. Journal of Hydrology, 291(3-4): 278-296.

Uhlenbrook, S. and Sieber, A., 2005. Sensitivity analyses of a distributed catchment model to verify the model structure. Journal of Hydrology, 310(1-4): 216-235.

Uhlenbrook, S. and Wenninger, J., 2006. Identification of flow pathways along hillslopes using electrical resistivity tomography (ERT). Symposium S7 held during seventh IAHS scientific Assembly. IAHS publication 303, Fez do Iguacu, Brazil, pp. 15-20.

URT, 1965. Geological survey of Tanzania, quarter degree sheet 89, Same, Geological division, Dodoma, Tanzania.

URT, 1977. Water Master Plan: Kilimanjaro Region, Ministry of Water, Energy and Minerals, Dar es Salaam.

URT, 2004. 2002 Population and Housing Census, Volume IV, District profile, Same, Central Census Office, National Bureau of Statistics, President's Office, Planning and Privatisation, Dar Es Salaam.

Valimba, P., 2004. Rainfall variability in southern Africa, its influences on streamflow variations and its relationships with climatic variations. PhD Thesis, Rhodes University, South Africa.

Van der Zaag, P., 2005. Integrated water resources management: relevant concept or irrelevant buzzword? A capacity building and research agenda for southern Africa. Physics and Chemistry of the Earth 30: 867-871.

Van der Zaag, P., 2007. Asymmetry and equity in water resources management; critical governance issues for Southern Africa. Water Resources Management, 21(12): 1993-2004.

Vrugt, J.A., Bouten, W., Gupta, H.V. and Soroshian, S., 2002. Toward improved identifiability of hydrologic model parameters: The information content of experimental data. Water Resources Research, 38(12): 1312.

Vyagusa, F.N., 2005. Water allocation practices for smallholder farmers in South Pare Mountains, Tanzania; A case study of Manoo Micro-dam (Ndiva). MSc Thesis, University of Zimbabwe, Harare.

Waalewijn, P., Wester, P. and Van Straaten, K., 2005. Transforming river basin management in south Africa; lessons from the Lower Komati river. Water International 30(2): 184-196.

Wels, C., Cornett, R.J. and Lazerte, B.D., 1991. Hydrograph separation: a comparison of geochemical and isotopic tracers. Journal of Hydrology, 122(1–4): 253–274.

Wenninger, J., Uhlenbrook, S., Lorentz, S. and Leibundgut, C., 2008. Identification of runoff generating processes using combined hydrometric, tracer and geophysical methods in a headwater catchment in South Africa. Hydrological Sciences Journal, 53(1): 65-80.

Wester, P., Merrey, D. and De Lange, M., 2003. Boundaries of consent: stakeholder participation in river basin management in Mexico and South Africa. World Development 31(5): 797-812.

White, W.N., 1932. A method of estimating groundwater supplies based on discharge by plants and evaporation from soil: Results of investigations in Escalante Valley, Utah.

Winsemius, H.C., Savenije, H.H.G., Gerrits, A.M.J., Zapreeva, E.A. and Klees, R., 2006. Comparison of two model approaches in the Zambezi river basin with regard to model reliability and identifiability. Hydrology and Earth System Sciences, 10: 339-352.

Winston, W.E. and Criss, R.E., 2002. Geochemical variations during flash flooding, Meramec River basin, May 2000. Journal of Hydrology, 265: 149-163.

SAMENVATTING

In verschillende delen van de wereld, en met name in sub-Sahara Afrika, zijn er veel stroomgebieden die nog nooit bemeten zijn. Informatie gewonnen in een bemeten stroomgebied en regionalisering van deze informatie naar onbemeten gebieden kan van cruciaal belang zijn voor het inschatten van de watervoorraaden in onbemeten gebieden. Met name de boeren in semi-aride gebieden hebben deze informatie hard nodig. De variatie in inter- en intraseizoensregenval is groot in deze gebieden. Omdat er steeds meer water gebruikt wordt en de variatie van de beschikbaarheid toeneemt, zijn boeren meer en meer afhankelijk van oppervlakte- en grondwatervoorraden om voedsel te verbouwen. Hierdoor is het begrijpen van de hydrologische processen en het bepalen van de frequentie en de grootte van de rivierafvoeren van belang voor de lokale voedselproduktie. Dit is in het bijzonder relevant voor het onbemeten Makanya stroomgebied in Tanzania, welke het onderwerp van deze studie is.

Aangezien er geen lange tijdseries van hydrologische gegevens beschikbaar zijn, zijn de hydrologische processen bestudeerd aan de hand van een "multi-method" benadering. Normale regen- en afvoermeetinstrumenten zijn hiertoe geïnstalleerd in een geneste opstelling. Ruimtelijke en temporele gegevens zijn met een hoge resolutie verzameld over een periode van 2 jaar. Watermonsters van lokale bronnen zijn verzameld om het grondwatersysteem in kaart te brengen. Het kwantificeren van de herkomst van het water en de stroompaden tijdens een hoogwatergolf zijn bepaald met behulp van de hydrograafscheidingsmethode gebruikmakend van hydro-chemische parameters. Electrische Weerstands Tomografie (ERT) is gebruikt om de ondergrondse structuur in kaart te brengen. Tenslotte is een conceptueel model ontwikkeld om de conceptuele veronderstelling van de stroompaden te toetsen.

Landbouwkundige praktijken in het stroomgebied verschillen per lokatie omdat deze worden beïnvloed door met name het lokale klimaat, de aanwezigheid van watervoorraden en grondsoort. In het studiegebied kunnen drie zones met verschillende karakteristieken worden onderscheiden. In het hoogland is het kouder, er valt meer regen en er bevinden zich door het hele jaar stromende (permanente) bronnen. In het regenseizoen wordt aanvullende irrigation toegepast in dit gebied, tijdens het droge seizoen volledige irrigatie. In de vallei wordt de overblijvende afvoer vanuit het hoogland gebruikt voor aanvullende irrigatie. In dit gebied is de waarschijnlijkheid van het voorkomen van een droge periode zo hoog dat het noodzakelijk is om aanvullende irrigatie te beoefenen. In dit gebied is er onvoldoende watertoevoer vanuit het hoogland om volledige irrigatie toe te passen. In het laagland is er tegenwoordig tijdens het droge seizoen geen basisafvoer meer vanuit de hooglanden en gebruiken de boeren alleen tijdens het regenseizoen vloedirrigatie, waarbij een vloedgolf wordt verdeeld over het landbouw gebied.

In het hoogland is de geologie bepalend voor het voorkomen van permanente bronnen. Deze bronnen genereren een grote hoeveelheid water. In de vallei komen er weinig permanente bronnen voor die per bron een substantiëel lagere hoeveelheid water van slechtere kwaliteit genereren. De overvloed aan bronnen in de hooglanden wordt zowel in het hoogland als in de vallei gebruikt voor irrigatie. Door het veelvuldige gebruik in deze gebieden wordt de infiltratie capaciteit van het rivierbed niet overschreden en bereiken de rivieren de uitgang van het stroomgebied niet meer. Alleen grote vloedgolven bereiken het irrigatie systeem in het laagland nog.

Er zijn twee verschillende vloedgolven geanalyseerd die door een verschillend type regenval werden veroorzaakt. Hoge intensiteit regenval genereert vloedgolven die het vloedirrigatiesysteem bereiken. Kleinere vloedgolven infiltreren in het alluvium, voordat ze het vloedirrigatiesysteem bereiken. Tijdens kleinere vloedgolven, zoals de vloedgolven van 9 november en 5 december 2005, blijft het overgrote deel van de hydrograaf vanuit het grondwater komen(>90%). Tijdens een extreme bui wordt de grondwatercontributie ongeveer 50% van de total afvoer. Hierbij wordt de groundwaterafvoer tot wel 100 keer verhoogd op het piekmoment en de basisafvoer volgend op de vloedgolf wordt ook substantieel verhoogd. De tijd tussen de regenbui en de piekafvoer is extreem kort, binnen één tot twee uur bereikt de piekafvoer de uitgang van het stroomgebied en stroomt het in het vloedirrigatiesysteem. De geanalyseerde regenval op 1 maart 2006, is zo'n extreme gebeurtenis, met regenval intensiteiten van 50 mm per uur, die maar 4 tot 5 uur aanhield. Tijdens deze regenbui overtrof de regen intensiteit de infiltratie capaciteit van de ondergrond, hierdoor kwam directe oppervlakteafvoer veelvuldig voor. De vloedgolf overtrof ook de absorptiecapaciteit van het rivierbed en vulde zo het vloedirrigatiesysteem aan.

De temporale resolutie van de meetgegevens is 15 minuten. Dit detailniveau was noodzakelijk om de snelle reactie van het stroomgebied tijdens piekafvoeren te kunnen waarnemen. Helaas zijn de meetgegevens niet compleet omdat tijdens extreme gebeurtenissen de meetinstrumenten werden beschadigd en de betrouwbaarheid van de gegevens aangetast, zoals op 1 maart 2006. Bovendien zijn 7 van de 10 instrumenten vermist, gestolen of beschadigd, wat het verzamelen van de gegevens ernstig heeft belemmerd.

Een conceptueel model is ontwikkeld om de hypotheses over de regenval-afvoer relatie te toetsen. Het model kan de afvoer aan de voet van het gebergte, tussen het hoogland en de vallei procesmatig goed simuleren. In het model zijn de hydrologische processen opgenomen. Het model geeft goede statistische resultaten voor een periode van 2 jaar met een tijdsinterval van één uur (Nash-Sutcliffe efficiëntie van 0.79 en Log Nash-Sutcliffe efficiëntie van 0.90). Het model bevat een grote hoeveelheid parameters, waarvan een aantal zijn bepaald aan de hand van de gegevens. Een automatische optimalisatie van de overige parameters wordt belemmerd door equifinaliteit. Daarom is er gekozen voor een stapsgewijze calibratie, waarbij gebruik gemaakt wordt van verschillende prestatiecriteria waarbij met name gekeken wordt naar verschillende onderdelen van de hydrograaf. Als dit model verder gebruikt zou gaan worden voor schaalvergroting van landbouwkundige praktijken in de vallei of het laagland, dan moet er rekening mee gehouden worden dat de hydrologische processen in de vallei van het stroomgebied anders zijn dan deze in het hoogland. Deze zullen verder bestudeerd moeten worden voordat het model toegepast kan worden in een groter gebied.

De huidige hydrologische processen en resulterende watervoorraden worden zoals in vele gebieden in Afrika sterk beïnvloed door de mens. Het toenemende watergebruik in de bovenstroomse gebieden in het stroomgebied heeft de noodzaak geschapen voor een overeenkomst tussen de watergebruikers in het hoogland en de vallei. Er bestaat echter geen overeenkomst tussen deze twee groepen en de boeren in het laagland. Aangezien de basisafvoer het dorp Makanya tegenwoordig niet meer bereikt, hebben de boeren een nieuwe irrigatie techniek moeten gaan toepassen. Veranderingen in het landgebruik in de bovenstroomse gebieden beïnvloedt ook de hydrologische processen daar. Maar aangezien historische gegevens ontbreken, is dit niet kwantificeerbaar. De benedenstroomse boeren kampen niet alleen met negatieve effecten. De vloedgolven die gegenereerd worden in het hoogland bereiken Makanya nu ook, en vullen daar de onverzadigde zone en het lokale grondwater aan. Hierbij worden ook vruchtbare sedimenten afgezet. Momenteel bestaat er een kwetsbare balans tussen de boven- en benedenstroomse boeren, waarbij in ieder gebied de mogelijkheid bestaat om voedsel te verbouwen. Het is echter onduidelijk hoe een toekomstige toename van de watervraag deze balans zou kunnen verstoren.

ABOUT THE AUTHOR

Marloes Mul was born on September 10, 1978 in Heinenoord, a small village on the island of Hoekse Waard, South Holland. In 2002, she obtained her MSc degree in Civil Engineering from Delft University of Technology, the Netherlands, with specialisation in Hydrology and Water Resources. After graduation she joined IHE-Delft, now UNESCO-IHE Institute for Water Education, as a lecturer in Water Resources Management.

In 2004, Marloes embarked on her PhD research at UNESCO-IHE. Between February 2004 and August 2007, she spent most of her time in Africa. She was hosted by the Civil Engineering department at the University of Zimbabwe. In the rainy seasons, Marloes went to do fieldwork in Makanya, a small village in Tanzania, setting up and maintaining monitoring equipment. Marloes was also involved in the management of the SSI research programme of which her PhD research is part.

During her PhD research, Marloes presented her work at several international conferences, which includes the Annual WaterNet/ WARFSA/ GWP-SA Symposia in Southern Africa, the European Geophysical Union (EGU), in Vienna, (April 2006) and UNESCO FRIEND Conference in Havana, Cuba (December 2006). She also supervised MSc students at the University of Zimbabwe, which generated three publications outside the scope of this research.

In January 2008, Marloes returned to UNESCO-IHE as lecturer in Water Resources Management, where she is now coordinating the MSc programme in Water Management. Along with these responsibilities, Marloes finalysed her PhD thesis in October 2008.

T - #0110 - 071024 - C66 - 254/178/8 - PB - 9780415549561 - Gloss Lamination